T0297090

Studies in Systems, Decision and Control

Volume 43

Series editor

Janusz Kacprzyk, Polish Academy of Sciences, Warsaw, Poland
e-mail: kacprzyk@ibspan.waw.pl

About this Series

The series "Studies in Systems, Decision and Control" (SSDC) covers both new developments and advances, as well as the state of the art, in the various areas of broadly perceived systems, decision making and control- quickly, up to date and with a high quality. The intent is to cover the theory, applications, and perspectives on the state of the art and future developments relevant to systems, decision making, control, complex processes and related areas, as embedded in the fields of engineering, computer science, physics, economics, social and life sciences, as well as the paradigms and methodologies behind them. The series contains monographs, textbooks, lecture notes and edited volumes in systems, decision making and control spanning the areas of Cyber-Physical Systems, Autonomous Systems, Sensor Networks, Control Systems, Energy Systems, Automotive Systems, Biological Systems, Vehicular Networking and Connected Vehicles, Aerospace Systems, Automation, Manufacturing, Smart Grids, Nonlinear Systems, Power Systems, Robotics, Social Systems, Economic Systems and other. Of particular value to both the contributors and the readership are the short publication timeframe and the world-wide distribution and exposure which enable both a wide and rapid dissemination of research output.

More information about this series at http://www.springer.com/series/13304

Hime Aguiar e Oliveira Junior

Evolutionary Global Optimization, Manifolds and Applications

 Springer

Hime Aguiar e Oliveira Junior
National Cinema Agency
Rio de Janeiro
Brazil

ISSN 2198-4182   ISSN 2198-4190   (electronic)
Studies in Systems, Decision and Control
ISBN 978-3-319-26466-0   ISBN 978-3-319-26467-7   (eBook)
DOI 10.1007/978-3-319-26467-7

Library of Congress Control Number: 2015954590

Springer Cham Heidelberg New York Dordrecht London

Printed on acid-free paper

Springer International Publishing AG Switzerland is part of Springer Science+Business Media
(www.springer.com)

Preface

Methods for solving global optimization problems with or without constraints find application in several disciplines like Game theory, Biology, Statistics, Engineering, Mathematics, Management Science, Economics, Physics, and in virtually every task that can be modeled by parametric systems. Many excellent methods have been developed so far, but several of them assume certain conditions about functions to be processed, as convexity or differentiability. However, when tackling real-world problems, we often get into situations in which objective functions to be optimized are not differentiable, convex, or even continuous. In such settings, many traditional methods are not so helpful, being necessary to find more general ways to get adequate results. In the past decades a large number of new global optimization methods were idealized aiming at reaching that more general objective, and a substantial part of them belong to the category of metaheuristic methods, being also known generically as metaheuristics. Many are based on probabilistic foundations, that is to say, use probability theory results to get to their objectives. Knowledge of the capabilities and limitations of these algorithms leads to a better understanding of their reach over various applications and indicates the way to future research on improving and extending algorithms' theoretical foundations and respective implementations. In another direction, almost all of them were developed to solve problems in linear spaces, not being able to deal with more general domains, like manifolds, for example. The main goal of this book is to present certain techniques for solving global optimization problems on manifolds by means of evolutionary algorithms, introducing certain applications chosen to complement the central presentation. In addition, the results will serve as a basis for constrained optimization in Euclidean spaces as well. By presenting detailed examples of applications we will try to stimulate the reader's intuition and make use of the proposed ideas easier to all wishing to solve constrained optimization problems on linear spaces or more general manifolds. In addition, applications tend to be concentrated on Game Theory, in particular Nash equilibrium problems related to several interesting real-world situations—they are reformulated as constrained global optimization problems and are solved with the help of Fuzzy ASA.

More abstract examples include minimization of well-known functions, in order to illustrate in detail the utilization of the proposed ideas. In order to offer usable material, the presented methods and examples use the Fuzzy ASA method, although many other paradigms could be adequate as well.

The insertion of an introductory chapter about metaheuristics is pertinent and justifiable because the algorithms presented in the book propose the use of existing evolutionary techniques when optimizing on manifolds. In this fashion, although Fuzzy ASA is used as the model optimizing paradigm, the suggested methods may be coupled to other metaheuristics. One big advantage of this viewpoint is that a great deal of tested knowledge may be applied in the new, expanded scope with almost null adaptation effort. The chapter also works as an indication and suggestion for future works, based on some of the fundamental ideas contained in the book.

Algorithms able to optimize functions defined on manifolds are important in that they may present lower computational complexity and frequently exhibit better numerical properties, as not getting caught in local minima attraction basins, for example. This becomes very clear when dealing with constrained optimization subject to equality constraints, given the inherent difficulty associated with such a type of restriction, due to its characteristic "bouncing" effect. This is so because it is not simple to keep evolving points exactly inside feasible regions using the conventional methods, giving rise to undesired oscillations along the optimization process. So, despite being important in more general contexts, even in traditional optimization tasks they may be quite effective. In this book it is presented a method for global optimization of functions defined on finite dimensional manifolds, which may be loosely described as configuration spaces that locally "look like" Euclidean spaces and, in truth, include them as particular cases, that is to say, $\mathbb{R}^n$ is a manifold as well. Pertinent elements of General and Differential Topology needed to develop the proposed algorithms are presented and it is possible to see that many already developed evolutionary paradigms can be applied almost directly, when faced and used in the adequate way. As many real-life problems can be naturally regarded as models whose defining parameters evolve on manifolds, like constrained optimization ones with equality constraints, for instance, new results in that direction are always welcome.

Prerequisites for reading this book include some knowledge of Linear Algebra, introductory Numerical Analysis and basic Probability Theory. Many necessary definitions and fundamental results are provided and formal mathematical requirements are kept to a minimum. The focus will be kept on continuous problems. This book can be used in courses related to optimization as well as by researchers and practitioners, and is adequate for self-study too.

The work is divided into three parts:

- Part I presents basic information about optimization algorithms, describing some well-known metaheuristics, their main characteristics, and overall architecture;
- Part II exposes fundamental facts about Topology, the Fuzzy ASA global optimization method along with its overall structure, well-known results about manifold theory, and the proposed methods themselves;

- Part III contains some important applications of (constrained) global optimization, with special emphasis on solutions of Generalized Nash equilibrium problems (GNEPs).

Rio de Janeiro
August 2015

Hime Aguiar e Oliveira Junior

Acknowledgments

I would like to thank Dr. Leontina Di Cecco (Editor in Engineeering/Applied Sciences), Prof. Janusz Kacprzyk (Series Editor), and all staff of Springer-Verlag for their kind support.

Contents

Part I
Introductory Information

Chapter 1
The Many Aspects of Global Optimization

Abstract This chapter contains several considerations about global optimization methods, exposing in a condensed way several of their main qualitative characteristics. Taking into account that the focus in this book will be on the combination of an evolutionary method along with results of Topology and corresponding applications, we'll start to pave the way that will take us to the practical utilization of that approach, and stochastic optimization methods, in general.

1.1 Introduction

Being global optimization a difficult area of study, the discovery of all global optima of multimodal functions (having more than one local extremum) is one major challenge in this field. The difficulty associated to this operation has led researchers to develop many techniques aimed at solving several difficult problems.

The efficacy of many of these algorithms is highly dependent on the set of starting points, not existing a definitive technique that may assure convergence to all global optimizers without being caught in suboptimal regions (containing local, nonglobal extremes).

For traditional, deterministic paradigms such as those known as gradient-based methods, which work using the calculation of the first or second derivatives of the cost function, it is not uncommon to be captured and converge prematurely. The convergence to global optimizers is often guaranteed only when members of the initial populations reside in the neighborhood of global solutions and objective functions obey certain (somewhat restrictive) conditions, such as convexity, for instance.

On the other hand, stochastic techniques such as some evolutionary algorithms are developed under diverse principles and motivations, and strive for overcoming the premature convergence problem, often by employing strategies of exploration and exploitation. Such procedures normally work by sweeping domains of cost functions, identifying probable regions containing global optimizers, and in the final phase, the search is concentrated in precisely locating the possible global solutions. In this manner, it is theoretically possible for such a type of methods to "travel" along different attraction basins, not getting stagnated inside suboptimal locations.

© Springer International Publishing Switzerland 2016

H. Aguiar e Oliveira Junior, *Evolutionary Global Optimization,*

Manifolds and Applications, Studies in Systems, Decision and Control 43,

DOI 10.1007/978-3-319-26467-7_1

Optimization methods have been a necessity since time immemorial even though people may not be conscious of this. As humans have always tried to reach optimal results, be them maximum personal income or minimum costs in industries, it is somewhat evident that optimization problems have always been around. Even in Nature, the protein folding process is supposed to search for a minimum energy configuration state for instance, and molecules forming solid bodies during the process of freezing tend to assume energy-minimum crystal configurations. Another good example is the biological principle of survival of the fittest [17] which, conjugated to species evolution theory [9], tries to model adaptation of the species to their environment, and is definitely related to optimization principles. Hence, it is no exaggeration to say that optimization problems pervade the Universe itself, where (usually nonlinear) optimization processes happen all the time at the microscopic, mesoscopic or macroscopic levels—in General Relativity, for example, traveling along geodesics on smooth manifolds is a fundamental premise. As optimization problems may be global or local, there are techniques aimed at finding local or global optima—a local minimization problem, for example, requires finding a minimizer within a proper, specific neighborhood. In this way, the problem is local, not encompassing the whole domain of the given objective function. On the other hand, global optimization methods handle the case in which the search is for optimum points over the whole domain of the objective function. Therefore, the goal of global optimization methods is to find the best elements from a set, possibly subject to a set of constraints—these conditions are expressed as mathematical functions or relationships. The objective function works as the index we want to optimize and the constraints are the relationships we want to satisfy [11, 13].

In general the objective functions are real valued, that is, their images are subsets of $\mathbb{R}$. Their domains may contain many kinds of elements, like numbers, vectors of real or integer numbers, for example. In another dimension, objective function values may be obtained not only by mathematical expressions, but as results of simulations that can, for example, involve long and computationally expensive operations, justifying to a certain extent why a large number of publications takes the number of cost function evaluations as a measure of efficiency.

1.2   Main Approaches to Optimization Problems

Classifying existing global optimization methods is not a simple task but, in general, they can be separated into two groups: deterministic and stochastic. The deterministic methods are most often used whenever objective functions present certain analytical characteristics that make it possible to employ known theoretical results so as to find optimal points [6, 16, 18]. When using gradient methods, for instance, there is frequently a simulation of a dynamical system that evolves towards local optima, according to decreasing directions of the function under processing. Therefore, the feasible space can be explored using another complementary technique, for instance, so as to accelerate the whole optimization task. On the other hand, whenever the

relationship between a solution candidate and its corresponding cost are too compli-
cated, it may be hard to solve the global optimization problem by means of deterministic methods, even when dealing with small dimension problems. In those cases,
stochastic algorithms can be a good alternative, particularly a specific and well-known
family of probabilistic algorithms: the so-called Monte Carlo-based approaches. Typically they are able to find solutions in reduced time but don't guarantee convergence
to global solutions. Their results might be non-global optima—when dealing with
multimodal objective functions even deterministic algorithms could not, in general,
ensure that they have reached global optima. In this fashion, we have an example
of a heuristic paradigm - in particular, the decision is taken with basis in previous
information and experimental data. Heuristics used in global optimization are useful
when choosing which one of a set of possible solutions is to be examined in the
sequence.

Heuristic approaches may be regarded (loosely speaking) as components of an
optimization algorithm that use information currently gathered by a given algorithm
to help in the decision of which candidate solution should be tested next or how
the next individual can be produced, for example. On the other hand, metaheuristics
are usually viewed as methods for solving general classes of problems, combining
objective function values and reasoning rules in a somewhat abstract and reasonably
effective way, often without exploring the structure of the particular problem at hand
[12, 14, 20]. This association is often performed stochastically by using samples from
the search space. Typical simulated annealing methods, for example, decide which
candidate will be evaluated next according to the Boltzmann probability distribution.
In another direction, biologically inspired algorithms try to follow the behavior of
natural evolution and consider solution candidates as individuals that compete in an
abstract setting. Considering other classification dimensions, metaheuristic methods
tend to be population-based, like genetic or particle swarm intelligence algorithms.
In addition to these nature-inspired and evolutionary approaches, there exist also
methods that try to behave similarly to physical processes: simulated annealing,
parallel tempering, and grenade explosion method [2, 12, 20].

1.3 Characteristics of Generic Global Optimization Problems

So far some global optimization algorithms were mentioned and we will have more to
say about them in the chapters ahead. However, the methods introduced in this book
will deal with only a small part of the actual number of available methods and we
intend to focus even more later. Nevertheless, it is a natural question to ask why there
are so many different algorithms and whether is this variety needed. One explanation
could be that there are so many different kinds of global optimization tasks, creating
different obstacles to optimizers and representing specific difficulties. Hence, in what
follows we discuss concisely the most common problems usually encountered by

existing paradigms during global optimization, namely, multimodality, stagnation, premature convergence, poor exploratory ability and related phenomena.

In practice, whenever a given implementation experiences even a single negative feature, as cited above, it can get "frozen" in a sub-optimal point and simply not be able to reach the global optimum. This is possible even if highly efficient optimization techniques are applied.

Figures 1.1, 1.2, 1.3, 1.4 and 1.5 show different types of configurations. As remarked before, all problems will be of the minimization type, and the graphs aim to illustrate the difficulties experienced when trying to obtain global minima starting from arbitrary points in the function's domain. Obviously the illustrations feature low dimensional cases, but when working in domains with higher dimensions there are similar scenarios.

To understand why these landscapes are hard to deal with, we should understand the situation in the optimization setting. One approach to obtain near-optimal solutions for complex problems in reasonable time is to apply metaheuristic optimization procedures—the fundamental fact is that optimization algorithms are guided primarily by values of objective functions. On the other hand, functions are considered hard from a numerical perspective if they are not continuous, not differentiable or

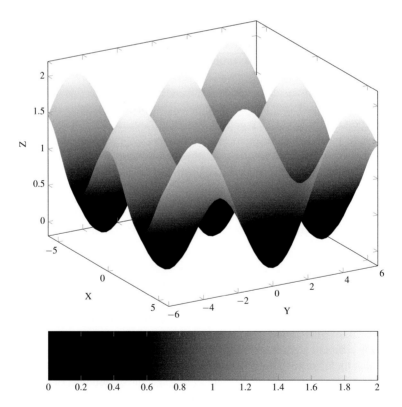

Fig. 1.1 Multimodal function

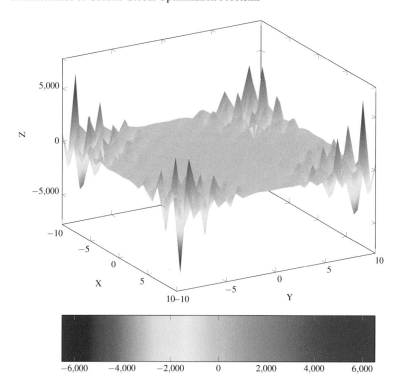

Fig. 1.2 Orientation is very difficult

multimodal. This conception of difficulty can be portrayed through the graphs shown in the figures mentioned above. In certain industrial applications of optimization techniques the analytical characteristics of the functions to be processed are not known in advance, and their values at different points are obtained from physical simulations or sampling operations. Hence it is seldom possible to predict the performance of optimizers when applied to these hard problems, in terms of having or not reached global optima, for instance. A reasonable and helpful strategy is to use models based on experimental data to improve the precision of final results.

   As a generic rule, it is assumed that an optimization algorithm has converged if it cannot produce new candidates or continues producing solution candidates located inside a very limited region of the problem domain. Another typical issue associated to global optimization is that it is often not possible to decide whether the best solution currently known is a local or global optimum and thus if a true solution has been found. Therefore, it is usually unclear when the optimization process should be interrupted, concentrate on refining the current optimum, or explore different parts of the search space. This is specially significant when dealing with multimodal and/or nondifferentiable objective functions.

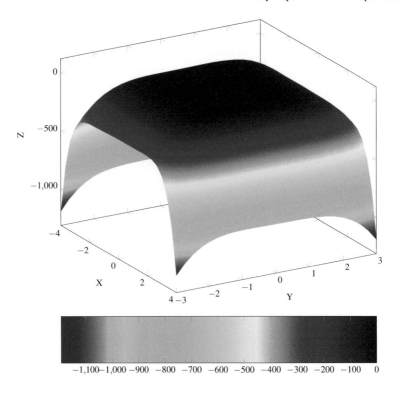

Fig. 1.3 Potential stagnation configuration

Premature convergence (to a local optimum) occurs if a given optimization process is no longer able to explore other parts of the search space except the area being currently visited (existing another region that contains a better solution) and stagnates at a sub-optimal point. Although the occurrence of multiple global optima is not very problematic, the existence of numerous local (and nonglobal) optima may originate problems, considering the possibility of premature convergence.

Yet another decision to be made is relative to the (exploration, exploitation) binomial. By exploration we mean visiting new sub-domains of the search space which have not been investigated earlier, and exploration processes try to find better solution states. Following this line, some operators are designed to create inferior solutions by destroying good individuals, but also have a small chance of finding more adequate candidates. On the other hand, exploitation processes incorporate small changes into existing individuals, leading to nearby solution candidates. Naturally, it is necessary to devise mechanisms that allow us to balance exploitation and exploration levels. It is worth to note that methods favoring exploitation over exploration have faster convergence but offer the risk of not finding the optimal solution, not forgetting the possibility of being caught at a local optimum. As expected, algorithms performing excessive exploration may never find the global optimum or spend too much

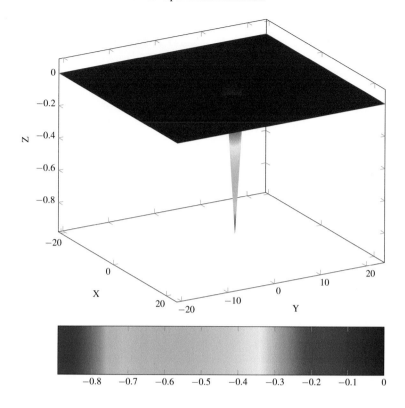

Fig. 1.4 Deep minimum

time trying to discover it. A good example might be the standard adaptive simulated annealing paradigm, that is able to do quenching, prioritizing exploitation over exploration and losing the guaranteed (theoretical) convergence in distribution to the global minimum.

Although there is no general method capable of avoiding premature convergence in all cases, the probability that an optimization process gets caught in a local optimum depends on several factors, including the properties of the function to be processed, for example. Typically, evolutionary global optimization methods adopt certain heuristics that tend to attenuate the risk of being caught in sub-optimal regions. As an example, increasing the degree of exploration may decrease the probability of premature convergence, taking into account that by doing so it is possible to improve the mapping of the objective function under study. As a matter of fact, some methods have been created to drive the search away from areas which have already been sampled. As an example, low selective pressure in a genetic algorithm decreases the chance of premature convergence but also delays the exploitation of good candidates. A possible alternative to avoid premature convergence is the use of self-adaptation, allowing the optimization algorithm to change strategies and its parameters, depending on the current state and performance figures. Such measures

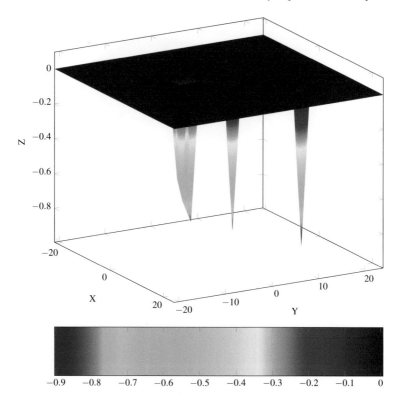

Fig. 1.5 Many needles

are often implemented in order to speed up the minimization process, even though the risk of converging prematurely to a suboptimal minimum may be raised.

Before closing this section, let us discuss a little about the situations represented in the previous graphs.

Figure 1.1 portrays a graph of a smooth cost function presenting several local minima and attraction basins, but only one global minimum in the given domain. Hence, depending on the starting point, algorithms will tend to converge to a non-global minimum. Therefore, a certain amount of preliminary exploration could be useful in terms of finding the best region to start the exploitation process.

Figure 1.2 displays a graph with strongly oscillatory and almost flat regions. Hence, a gradient-based algorithm would have problems in finding global minima when applied to this function because, starting from the almost constant part it will converge to a nearby local minimum, whereas initiating from more oscillating regions, those methods would probably arrive to a non-global minimum. In this fashion, methods with stronger exploratory behavior are likely to perform better.

In Fig. 1.3 it is possible to see another configuration with a small region presenting no variation. In such a case, starting points located there will not move because their evolution is usually based on increments proportional to gradient values. Whenever initial points are located in the peripheral regions, typical gradient-based methods will tend to perform well, finding global optima.

Another tough situation is shown in Fig. 1.4, with almost flat peripheral regions where gradient-based algorithms would probably get stagnated. A small central region contains the global minimum.

Finally, Fig. 1.5 represents a nightmare in terms of global optimization, considering that the function to be processed has three deep minima separated by almost plane regions inside which derivatives are not so useful for indicating good directions to follow.

Although there are many global optimization methods, only a few have gained high popularity. For example, in the differentiable realm there are the gradient-based ones and their variations [3–5, 18], whose orientation is based on the negative of cost functions' gradient (for minimization problems). That is to say, departing from a chosen point, these methods usually evolve by combining the current state to terms that tend to drive the search toward regions in which the objective function assumes lower values. The net result is often satisfactory inside attraction basins because the gradient at a given point gives information about the variation tendency of the function in a neighborhood of that point but whenever the seeds are located in inadequate regions the search is taken to a suboptimal location. We can also highlight partition algorithms [19] for continuous global optimization problems—these methods are very effective and adequate for minimizing continuous or Lipschitz-continuous objective functions over compact domains in $\mathbb{R}^n$.

Fortunately, there are stochastic and metaheuristic approaches that present good results with difficult optimization problems. As examples we have: simulated annealing, genetic algorithms, cross-entropy, artificial bee colony, particle swarm optimization, grenade explosion method, differential evolution, and further nature-inspired methods [1, 7, 8, 10, 14, 15, 17, 20]. These methods have been useful when applied to many areas of knowledge, and they represent excellent tools for solving several nontrivial real world problems.

This can be explained thanks to their ability to escape from local minima and flexibility in accepting transitory degradations during a certain time span in order to find better states in latter phases of optimization processes.

References

1. Aguiar e Oliveira Jr, H., Ingber, L., Petraglia, A., Petraglia, M.R., Machado, M.A.S.: Stochastic Global Optimization and Its Applications with Fuzzy Adaptive Simulated Annealing. Springer, Berlin (2012)
2. Ahrari, A., Atai, A.A.: Grenade explosion method—a novel tool for optimization of multimodal functions. Appl. Soft Comput. **10**, 1132–1140 (2010)
3. Bazaraa, M., Sherali, H., Shetty, C.: Nonlinear Programming, Theory and Applications. Wiley, New York (1993)

4. Bertsekas, D.P.: Constrained Optimization and Lagrange Multiplier Methods. Academic Press, New York (1982)
5. Bertsekas, D.P.: Nonlinear Programming. Athena Scientific, Belmont (1995)
6. Brent, R.P.: Algorithms for Minimization Without Derivatives. Prentice Hall, Englewood Cliffs (1973)
7. Cherruault, Y., Mora, G.: Optimisation Globale—Theorie Des Courbes Alpha-Denses. Economica, Paris (2005)
8. Clerc, M.: Particle Swarm Optimization. ISTE Publishing Company, London (2006)
9. Deb, K.: Multi-Objective Optimization Using Evolutionary Algorithms. Wiley, New York (2001)
10. Dorigo, M., Stützle, T.: Ant Colony Optimization. Bradford Books, Cambridge (2004)
11. Gill, P., Murray, W., Wright, M.H.: Practical Optimization. Academic Press, London (1981)
12. Hartmann, A.K., Riger, H.: Optimization Algorithms in Physics. Wiley, Berlin (2002)
13. Himmelblau, D.M.: Applied Nonlinear Programming. McGraw-Hill, New York (1972)
14. Karaboga, D., Basturk, B.: A powerful and efficient algorithm for numerical function optimization: artificial bee colony (ABC) algorithm. J. Glob. Optim. **39**, 459–471 (2007)
15. Karaboga, D., Basturk, B.: On the performance of artificial bee colony (ABC) algorithm. Appl. Soft Comput. **8**, 687–697 (2008)
16. Lawson, C.L., Hanson, R.J.: Solving Least Squares Problems. Prentice-Hall, Englewood Cliffs (1974)
17. Michalewicz, Z.: Genetic Algorithms + Data Structures = Evolution Programs. Springer, Heidelberg (1994)
18. Nocedal, J., Wright, S.J.: Numerical Optimization. Springer, New York (1999)
19. Pintér, J.D.: Global Optimization in Action. Kluwer Academic Publishers, Dordrecht (1996)
20. Weise, T.: Global Optimization Algorithms—Theory and Application. Available as e-book at http://www.it-weise.de/. Accessed 11 July 2011

Chapter 2
Overview of Current Metaheuristic Paradigms

Abstract At this point some of the most visible paradigms related to global optimization using metaheuristics are described. The general intention is to describe well-established methods and show their usefulness in several difficult optimization scenarios. Although there exist a reasonable number of effective algorithms in this class, only a few of them are presented here, in the hope that the selection may give the reader a good idea of the whole context. In principle each one may be used as an "optimization engine" when doing global optimization on manifolds, to be described later.

2.1 Introduction

Many problems presenting high complexity can often be expressed as optimization tasks, frequently global optimization ones. Hence, it is usually necessary to define an objective function so that its extreme points with respect to a set of parameters represent the solution to the original question. Frequently the optimization problem is imposed a constraint set [1, 15, 20]. The aim of this chapter is to concentrate on a group of methods, namely metaheuristics, that include paradigms like differential evolution, simulated annealing, genetic algorithms, PSO etc. Having the same common aim of solving hard optimization problems, their working principles are typically inspired by physical, pre-existing systems [2–4, 6–8].

Sometimes the metaheuristics label is not seen as an adequate expression, often used to describe a subfield of stochastic optimization, that is the general class of techniques employing randomness to search for optimal solutions. Metaheuristics could be considered the most general of these types of algorithms, being applied to a very wide range of problems [9].

In this fashion, metaheuristics methods are, to a large extent, based on a common set of principles that make it possible to design powerful algorithms. The large variety of existing metaheuristics could be attributed to different ways of combining those fundamental ideas.

Perhaps due to this fact, it is a not so uncommon mistake to confuse things and say that different paradigms are the same only because they use exploration and exploitation in their search mechanisms.

© Springer International Publishing Switzerland 2016
H. Aguiar e Oliveira Junior, *Evolutionary Global Optimization,*
Manifolds and Applications, Studies in Systems, Decision and Control 43,
DOI 10.1007/978-3-319-26467-7_2

Of course, almost all stochastic global optimization algorithms do use exploration and exploitation in a probabilistic way, including nature-inspired ones. But this "logic" is wrong, taking into account that the most effective algorithms implement their heuristics in different ways, obviously directed toward maximum efficiency and effectiveness. Anyway, it is advisable to devise ways of classifying metaheuristics.

It is not difficult to find the categories into which metaheuristic methods may be classified. Depending on the chosen criteria, many subdivisions are possible, each of which is the result of a specific perspective. In this chapter the most common ways of classifying metaheuristics are described.

It is common practice to base the classification on the fundamental "inspiration" of the algorithm, as in the division based on whether they are nature-inspired or not nature-inspired ones. Hence, examples of nature-inspired algorithms are particle swarm optimization (PSO) and artificial bee colony (ABC) methods [11, 12, 16], and not nature-inspired ones could be grenade explosion and cross-entropy. Naturally, there are occasions when it is not simple to classify a given paradigm, but it is somewhat usual to find references to this classification scheme in the literature.

A different classification scheme is related to the number of states stored by algorithms during their search processes, namely, population-based and single point search. This scheme is based on the number of candidate solutions managed simultaneously and, as the name indicates, depends on whether the algorithm keeps one or many candidate solutions during the search process. Single solution methods tend to be simpler in terms of data management and almost memoryless, describing a trajectory in the search space during the whole optimization process. Population-based methods, on the other hand, drive multistate processes, evolving many auxiliary components along the whole search period.

Yet another way to classify metaheuristics is to consider the way they use objective functions. Most algorithms use a fixed objective function, while others modify it during the search, trying to escape from local minima by changing the original landscape. Therefore, during the optimization process the objective function can be modified, depending on the information obtained so far.

In the following, the main characteristics of some well-known methods are described.

2.2 Particle Swarm Optimization (PSO)

The PSO paradigm belongs to the class of swarm intelligence methods, common approach when solving global optimization problems [4, 16]. Used at first to simulate social behavior scenarios, it was employed as an optimization method in the nineties. Being related to swarm-based theories and evolutionary computing, it is easily implemented in well-known programming languages and demands few computational resources. Furthermore, it requires only objective function values (no derivative information), demonstrating to be an efficient method for complex optimization problems. Its fundamental working principle is based on maintaining a population of candidate solutions to the problem under study, whereas each individual in the population has variable velocity, making it move through the state space. In addition each particle possess a private memory, recording the best position of the search space that it has ever visited so far in the progressing simulation. In this fashion, the evolution proceeds toward a collective tendency to the best previously visited positions.

Showing coherence with the fundamental working principles of swarm intelligence heuristics, its basic operations are simple and inexpensive in computational terms, being also responsive to changes in the overall environment, not changing its behavior every time the context changes.

It is noticeable that there are differences between PSO and usual evolutionary computing techniques in that the latter feature typically three main operations, namely, recombination, mutation and selection. PSO does not present a strict recombination operation, but the stochastic acceleration of an individual toward its previous best position, as well as in the direction of the best particle of the group, is similar to the recombination procedure in evolutionary processes. Also, the information swap occurs only between the particle's own history and that of the best particle in the set, instead of being carried from fitness dependent selected "parents" to "descendants". Furthermore, the directional position updating operation of PSO could be compared to the mutation of GA.

In summary, PSO happens to be a useful technique for solving complex global optimization problems, although additional research is necessary to completely establish its dynamics and full power.

2.3 Differential Evolution (DE)

It is a simple, easy to use, and effective method for global optimization of real-valued, multimodal objective functions [18]. Its implementations are typically efficient, not requiring too much housekeeping, with respect to parameter tuning. In its standard construction, it utilizes just a few control variables which are kept fixed during optimization procedures, but there are many variants of differential evolution with adaptive dynamics. In this optimization model, populations are formed by sets of individuals represented by vectors. New parameter vectors are synthesized by adding the weighted difference between two population vectors to a third vector. When the resulting vector is better than a given individual, a substitution takes place.

Although the method is relatively new, there exist several variants that characterize themselves by the driving strategy they modify/evolve existing vectors, that is to say, it is possible to compose individuals with more than one weighted difference vector or mix the parameters of the old vector with those of the perturbed one before comparisons. The many existing schemes normally differ in the way a new element is created—operations similar to GA's crossover occur after generating new points, by composing certain aspects of involved individuals. To make the actual combination, an offset within the vector and a crossover amplitude are randomly chosen, being the control parameter for the crossover size a number in the interval [0, 1], determining its probability distribution. This crossover-like procedure works by swapping a strip of a certain size from the generated individual and the corresponding part of the current one. The next step is to evaluate the resulting point, replacing the present one in case of improvement.

The basic set of parameters controlling the dynamics of this method consists of population size, "crossover-like" amplitude and a non-null element of $[-1, 1]$, the latter controls the degree of influence that differences have on the generation of mutant individuals. After this brief description it is possible to conclude that differential evolution is very easy to deal with and apply, but there is no consensus of which scheme is the most adequate, nor is there a definitive indication of which parameter values to use in each case. Nevertheless, this paradigm keeps improving since its original version and have been applied to several complex problems, demonstrating very good performance in many of them.

2.4 Genetic Algorithms

This paradigm was introduced decades ago and is still intensively used today, especially its more recent variations, incorporating increasing degrees of adaptability [10, 14]. The fundamental algorithm is very simple and works by iterating through fitness evaluation (fitness function calculation), selection, mating and population re-arrangement. The differences among the several implementations reside, for example, in the way selection and breeding occurs. In addition, new ways of chromosome coding may provoke a significant difference in the final algorithm—typically it is possible to find binary and floating-point types. They deeply differ in the way operations of mutation, reproduction and recombination are constructed. Actually, many meanings are attributed to the term *genetic algorithm*, but a genetic algorithm may be defined as a population-based method using selection, reproduction and mutation operations, aiming to evolve a variable set of candidates in a certain state space toward a point of maximum fitness. Since the initial formulation, several types of genetic operators and chromossome representations were proposed for solving a large number of optimization problems. Being inspired by biological evolution, the encoding of potential solutions uses chromosome-like data structures that are submitted to recombination operations so as to preserve positive features. The initial population is composed of individuals (chromosomes) usually chosen in a randomized way and. from then on, evolution occurs by evaluating elements in each generation and inducing reproduction so that better solutions have greater probability of generating offspring. The classical procedure for evolving a given population begins with an empty set, selects parents from the original population, then copies, combines them with one another, mutating the resulting individuals—this sequence happens until a new population is formed.

As expected, more adequate individuals are given greater opportunities to reproduce and the genetic operators (usually mutation and crossover) are applied to the individuals in the mating buffer, producing offspring that will compose the next generation. Obviously, mutation and crossover processes can be designed in several ways, in particular the mutation rate is a very influential parameter, considering that if the rates are low, the resulting offspring tend to be very similar to their parents, and vice versa. In a distinct design dimension, it is necessary to decide about how many individuals are to be produced by crossover, and how many are to be selected and paired in the reproduction buffer. As almost all genetic algorithms keep fixed-sized populations, it is necessary to select the given number of individuals resulting from the overall processes described above. It is possible, for instance, to keep all generated children and select individuals from the old population to survive in the new population. In that case, each "birth" in the new population implies in one "death" in the old population, so to speak. In the limit (two new individuals associated to two parents), the previous population completely vanishes, being totally replaced by the new population. If reproduction is based on the fitness degree, the probability of an individual being chosen is mainly based on its objective function value. This strategy may be implemented by scaling fitness values of the parent population and computing the probability of any individual being selected, with basis on the respective values. Although scaling mechanisms are able to attenuate the problem of weak selective pressure, not rarely the use of fitness-based selection may have the opposite effect, that is, increased selective pressure. Therefore, individuals presenting high fitness have greater tendency of being selected. Going further, if many instances of a high fitness individual are placed in the mating buffer, the new population will probably contain many of its clones, provoking premature convergence. To avoid this, fitness proportional selection is replaced with ranked selection, that ranks individuals in the parent population, making the probability of selection a function of rank instead of fitness. Another well-known way of performing choices is tournament selection, in which a reduced subset of individuals is randomly chosen and the best individuals are selected for reproducing. A common type of selective criterion that preserves copies of the best individuals is referred to as elitism—it is an attempt to preserve the current quality, expecting that it could help finding better elements in future generations.

2.5 Grenade Explosion Method (GEM)

The underlying idea of the grenade explosion algorithm [2] is (obviously) inspired by the dynamics of a grenade explosion, during which the thrown pieces of shrapnel destroy objects near the location where the explosion takes place. The damage caused by fragments of shrapnel is computed, and high values for loss per piece of shrapnel in an area indicates the existence of valuable goods in that neighborhood. In order to provoke more losses, the next grenade should be directed to the point with greatest losses. Even though objects nearby grenade's location are more likely to be destroyed, the chance of destruction for distant objects may be amplified by choosing a high value for the parameter representing the length of explosion along each coordinate. By repeating this procedure, it would eventually result in finding the best place for throwing the grenades. The fitness of the objective function at the object's location is taken as the loss caused by destruction of an object.

One outstanding feature of this paradigm is the agent's territory radius, representing the idea that agents (the same as grenades) do not permit other agents to come closer than a specific distance. Hence, whenever several agents explore the feasible space, higher values for this parameter force grenades to be positioned in a more uniform way in the search domain and the whole space tends to be explored. On the other hand, lower values allow grenades to get closer in order to investigate local regions jointly. Better exploration of distant regions is achieved when higher values for the explosion range are established, whereas lower values tend to drive the algorithm to focus the search on smaller sub-domains, leading to better exploitation.

By comparing results obtained from the optimization of several benchmark functions using GEM [2] and other evolutionary methods, it is evidenced that GEM is able to find global minima locations efficiently. Besides, according to the tests, it is often able to find the global minimum of multimodal functions without being caught inside local minima attraction regions. In summary, the method has shown promising performance but may be improved—this can be achieved, for example, by automatically tuning certain control parameters or pruning the overall control set, in order to reduce the housekeeping effort.

2.6 Artificial Bee Colony Algorithm (ABC)

ABC algorithm [11] is population-based and maintains a colony of artificial agents (bees) that can be of three types: employed, onlookers, and scouts. Onlookers are bees waiting in the dance area for making decisions about food sources. Employed bees go to food sources visited by themselves previously. Scouts execute random searches. The first half of the colony is integrated by employed bees and the second half by onlookers, and to each food source corresponds only one employed bee. Employed bees become scouts whenever their food sources get exhausted.

Loosely speaking, after initializing data structures, the basic algorithm works by cycling through three steps:

- Directing employed bees to food sources and measuring the respective nectar amounts;
- Selecting food sources by onlookers after sharing the information of employed bees and determining nectar amount of the foods;
- Choosing the scouts and sending them to potential food sources.

During the initialization stage, a collection of food source positions are randomly chosen by bees and their nectar amounts are calculated. After this, bees get into the hive and share the nectar information with other ones waiting in the dance area, inside the hive. In the course of the second stage, the information is shared and all employed bees go to the food source area visited by themselves in the previous cycle, then choose new food sources using visual information in the surroundings. In the third phase, onlookers favor food sources areas depending on nectar information spread by the employed bees in the dance area. In this fashion, the probability of choosing a food source is directly proportional to the nectar amount existing in it. So, onlookers for the food source areas with higher nectar amount are selected and, after arriving at the determined area, a new food source is chosen in the neighborhood of the one in the memory depending on visual information, which is based on the comparison of food source positions. Also, whenever the nectar of a food source is abandoned by bees, a new food source is determined by a scout bee and occupies the position of the discarded one. In the original method, at each cycle, only one scout searches for new food sources, and the same number of employed and onlooker bees is used. In the original algorithm, positions of food sources represent possible solutions for the optimization problem, and nectar quantities associated to food sources correspond to quality (fitness) of corresponding solutions. Furthermore, the number of employed bees is the number of solutions and, firstly, ABC generates a random initial population. Each food source is represented by a vector containing the parameters under optimization, and the population of potential solutions suffers successive transformations through many cycles of the search processes of the employed, onlooker, and scout bees. Employed or onlooker bees produce random modifications to solutions in the memory for the sake of finding new food sources and sample the nectar amount of the new source. Similar to what occurs in Nature, where the selection of new food sources is based on comparisons among food sources in a certain region, depending on the available information, in the ABC paradigm the detection of new food sources is also based on comparison of food source positions. Differently from the real ones, artificial bees do not use any information when making comparisons, they just select food source positions at random and produce modifications in their memories. In addition, if the nectar content of the new source is greater than that of the previous one, the new position is stored and the old one forgotten. In the contrary case, the position of the previous one is maintained and, after all employed bees complete the search process, the nectar and position information of food sources is shared with the onlooker bees. Onlookers evaluate nectar information obtained from employed bees and choose food sources in a random way, but related to their content. As in the case of employed bees, a change in positions is produced in memory and the nectar amounts of candidates are checked. If the nectar quantity is higher than that of the previous one, the new position is kept and the old one forgotten. Considering that in robust search processes exploration and exploitation mechanisms are very important and complementary, in ABC, onlooker and employed bees are responsible for executing the exploitation task, while scouts control the exploration of the search domain.

In summary, in [11] ABC is compared to GA, PSO, and PS-EA, which are swarm and population based approaches as well . The tests demonstrated good performance of ABC algorithm when optimizing five high dimensional cost functions featuring multimodal landscapes, and the conclusion is that it is able to escape from local minima and can be efficiently used for multivariable, multimodal function optimization.

2.7 Cross-Entropy Method (CE)

This algorithm was introduced in 1997, aimed at estimating probabilities of rare events by minimization of sample variance of the importance sampling estimator. Later modified to minimize the Kullback-Leibler divergence, evolved into the present cross-entropy method [17]. The CE method can be easily applied to the solution of continuous multimodal optimization problems [17] and may be described as an iterative algorithm which consists of repeating two steps in each execution cycle, namely, generation of random data samples using a set of dynamic parameters and updating the current set of parameters controlling the generation of random data using the current sample, trying to improve data in the next iteration.

The cross-entropy method presents a versatile structure, easily adjustable to a wide range of applications, with a very simple and efficient updating procedure. The measurements (individuals) are obtained from simulations as a random vector $X = (X_1, \ldots, X_n)$ having probability density function f. As in other types of evolutionary algorithms, information about the problem is contained in a population of potential solutions. In CE, the information about the experiment is concentrated in the current pdf, narrowing down the search for information by compressing it into a parametric vector. The accuracy with which the true parameter vector can be estimated from a result of a given simulation is related to the amount of information about it which is contained in the present sample X.

Another important feature is the choice of a method for measuring the amount of information by using the concept of distance which, in CE, is the Kullback-Leibler divergence. On the other hand, the measurement of information may be done by using the concept of Shannon entropy, which defines the average number of bits required to send a message X through a communication channel.

As mentioned above, the paradigm has two phases, which are repeated until a pre-defined stopping criterion is met: after producing the initial sample according to a specified probabilistic mechanism, the relevant parameters are modified with basis on the current population. Following this, a new sampling phase starts (using the newest estimated parameters) in order to obtain a better estimative of the true parameters, corresponding to the actual population. This process repeats until the solution is obtained or the maximum number of iterations is reached.

Therefore, the CE method may be considered as a kind of search process based on importance sampling distributions that approximates the true distribution corresponding to the objective function under minimization and, to reach that aim, Kullback-Leibler divergence is taken as a kind of proximity measure, guiding the overall approximation task. Hence, it can be viewed as a minimization problem whose search space is a class of probability distribution functions, and the objective function is proportional to the difference between candidate probability distribution functions and the function describing the actual landscape.

2.8 Simulated Annealing

Initially designed to handle combinatorial optimization problems [13, 19], this method has been successful in high-dimensional domain problems, being based on evaluations of objective functions, and allowing transitions out of sub-optimal attraction basins. Although there is no guarantee of finding global minima, the algorithm tends to approach them in a probabilistic sense. It is also able to discriminate between large oscillations of the landscape and small perturbations. The guiding principle is, at first, to exploit objective function's domain and reach areas in which at least one global minimum must be present. At this stage, it tries to find good, near-optimal local minima, or perhaps a global minimum. In [5] some modifications to this algorithm are made in order to use it in the optimization of functions defined on continuous domains—these functions do not need to be smooth or continuous in their domain, and the algorithm works based exclusively on objective function evaluations. The method assumes that $f : C \to \mathbb{R}^n$ is the bounded function to minimize, where C is a hyper-rectangle and f does not need to be continuous. It starts from a given point x_0 and generates a succession of points $x_1, x_2, \ldots$ approaching the global minimum of the cost function f. During each new iteration, candidate points are generated around the current point x_i by perturbing it along each coordinate direction.

The SA algorithm starts at a user defined temperature T_0 and generates a sequence of points until statistical mixing occurs. Next, the temperature T is lowered and a new sequence of moves is made, starting from the best point resulting from the previous phase, until a new equilibrium takes place again. The process is stopped at a temperature such that no more significant improvement could be reached, satisfying a pre-established criterion.

The simulated annealing framework may be viewed as a method of simulation of physical processes in which metallic objects are driven to states of minimum energy. In Nature this is possible by cooling metals slowly, until they get to a highly ordered state of low energy levels. On the other extreme, fast cooling schedules provoke a kind of disorder inside the material, making it harder after reaching lower temperatures. This behavior is compatible with the experimental fact that a search process that only accepts new points with lower cost function values tends to get stuck in local minima attraction basins. In this fashion, due to its decision criterion it permits uphill moves. When at higher temperatures, only the global behavior of the cost function is relevant to the search dynamics and, as temperature values decrease, smaller neighborhoods can be explored, allowing the algorithm to reach more precise results. Not existing deterministic guarantees of reaching global minima, there is a statistical proof of convergence and, in practice, the method is typically able to proceed toward global minima even in the presence of multimodal landscapes.

Although it is expensive in terms of number of function evaluations, it is capable of finding global minima of difficult objective functions and provides a very high reliability in the minimization of multimodal functions. In another direction, some design decisions have to be addressed as, for example, the choice of starting temperatures. By choosing a very high one, there will be a waste of computational resources.

In the opposite extreme, it is very likely to be caught in sub-optimal neighborhoods. In addition, objective functions' landscapes vary to such an extent that it is very hard to establish general rules for calculating good starting temperatures and, in practice, some experimentation may be of great help in finding adequate values for it and other significant control parameters.

While at high temperatures, practically any change is accepted—the algorithm sweeps a relatively large territory around the current state. On the other hand, at lower temperatures, transitions to higher energy states become less frequent and the search tends to concentrate in smaller regions. The cooling schedule is very important to the performance of simulated annealing approach, consisting of initial temperature, decrement function for the temperature, final temperature determined by a stopping criterion, and the number of transitions in the homogeneous Markov chain at each temperature stage. As expected, the quality of the final solution is directly proportional to the total execution time, the latter being dependent on the decrement speed of the temperature.

References

1. Aguiar e Oliveira Jr, H., Ingber, L., Petraglia, A., Petraglia, M.R., Machado, M.A.S.: Stochastic Global Optimization and Its Applications with Fuzzy Adaptive Simulated Annealing. Springer, Berlin (2012)
2. Ahrari, A., Atai, A.A.: Grenade explosion method—a novel tool for optimization of multimodal functions. Appl. Soft Comput. **10**, 1132–1140 (2010)
3. Birbil, S.I., Fang, S.: An electromagnetism-like mechanism for global optimization. J. Glob. Optim. **25**, 263–282 (2003)
4. Clerc, M.: Particle Swarm Optimization. ISTE Publishing Company, London (2006)
5. Corana, A., Marchesi, M., Martini, C., Ridella, S.: Minimizing multimodal functions of continuous variables with the simulated annealing algorithm. ACM Trans. Math. Softw. **13**, 262–280 (1987)
6. Deb, K.: Multi-Objective Optimization Using Evolutionary Algorithms. Wiley, New York (2001)
7. Dorigo, M., Stützle, T.: Ant Colony Optimization. Bradford Books, Cambridge (2004)
8. Dréo, J., Pétrowski, A., Siarry, P., Taillard, E.: Metaheuristics for Hard Optimization Methods and Case Studies—Simulated Annealing, Tabu Search, Evolutionary and Genetic Algorithms, Ant Colonies. Springer, Berlin (2006)
9. Glover, F., Kochenberger, G.A.: Handbook of Metaheuristics. Springer, Boston (2003)
10. Holland, J.H.: Adaptation in Natural and Artificial Systems. University of Michigan Press, Ann Arbor (1975)
11. Karaboga, D., Basturk, B.: A powerful and efficient algorithm for numerical function optimization: artificial bee colony (ABC) algorithm. J. Glob. Optim. **39**, 459–471 (2007)
12. Karaboga, D., Basturk, B.: On the performance of artificial bee colony (ABC) algorithm. Appl. Soft Comput. **8**, 687–697 (2008)
13. Kirkpatrick, S., Gelatt Jr, C.D., Vecchi, M.P.: Optimization by simulated annealing. Science **220**, 671–680 (1983)
14. Michalewicz, Z.: Genetic Algorithms + Data Structures = Evolution Programs. Springer, Heidelberg (1994)
15. Nocedal, J., Wright, S.J.: Numerical Optimization. Springer, New York (1999)

16. Parsopoulos, K.E., Vrahatis, M.N.: Recent approaches to global optimization problems through particle swarm optimization. Nat. Comput. **1**, 235–306 (2002)
17. Rubinstein, R.Y., Kroese, D.P.: The Cross-entropy Method: A Unified Approach to Combinatorial Optimization, Monte-Carlo Simulation, and Machine Learning. Springer, New York (2004)
18. Storn, R., Price, K.: Differential evolution—a simple and efficient heuristic for global optimization over continuous spaces. J. Glob. Optim. **11**(4), 341–359 (1997)
19. van Laarhoven, P.J.M., Aarts, E.H.L.: Simulated Annealing: Theory and Applications. D. Reidel, Dordrecht (1987)
20. Weise, T.: Global Optimization Algorithms—Theory and Application. Available as e-book at http://www.it-weise.de/. Accessed 11 July 2014

Part II
Global Optimization on Manifolds

Chapter 3
Evolutionary Global Optimization on Manifolds

Abstract In this chapter it is described an approach to globally optimize real valued functions defined on topological manifolds. The functions under investigation do not need to be differentiable or even continuous. It is shown that optimization processes may take place so that candidate points remain restricted to the manifolds that contain their domains—the evolution occurs inside them during the entire optimization process. The presented paradigm is adequate for use with virtually all already existing metaheuristics, but here the algorithm known as Fuzzy Adaptive Simulated Annealing (Fuzzy ASA) is used in order to exemplify the overall scheme. After exposing the fundamental ideas, some examples will illustrate the efficacy of the proposed method.

3.1 Introduction

Although global optimization on Euclidean spaces is a very mature and well-established area of research, techniques for dealing with the same problem on curved spaces (manifolds) are still rare and practically restricted to gradient related methods. In this chapter, an extension of this setting is established by applying stochastic/evolutionary methods to the optimization of objective functions defined on manifolds. Several recent research efforts involving optimization on matrix manifolds [1] have provided significant alternatives to many constrained optimization methods. Optimization algorithms capable of working with cost functions defined on manifolds may present lower computational cost and often may also have better numerical properties, escaping from local minima attraction basins.

The present chapter deals with a proposal for global optimization of functions defined on finite dimensional manifolds, which can be loosely described as state spaces that are locally similar to (finite dimensional) vector spaces. So, after describing certain basic elements of General and Differential Topology and explaining the main ideas, it will be possible to see that many already developed paradigms can be applied almost directly, when used in the proper way. By observing that many real world problems may be regarded as models whose parameters naturally evolve on

© Springer International Publishing Switzerland 2016
H. Aguiar e Oliveira Junior, *Evolutionary Global Optimization,*
Manifolds and Applications, Studies in Systems, Decision and Control 43,
DOI 10.1007/978-3-319-26467-7_3

manifolds, like constrained optimization ones with equality constraints, for instance, it is possible to realize how important it is to obtain new methods in that direction.

The generic method here described can be directed to several optimization tasks, defined only over certain regions and modeled as topological manifolds. Nowadays there is a large amount of results about more restricted classes of problems, generally related to matrix optimization tasks, and characterized by symmetry or invariance properties in the objective function or associated constraints. Those topics abound in algorithmic issues related to several subjects [12], in general using specific characteristics of each context in order to improve final results. The proposed idea here is different, considering that the application of the same basic (evolutionary) algorithms is prescribed, changing only the specification of the objective function. Generally the strategy is effective, if compared to the cited above. To furnish some examples of past research works, in [1] several techniques for solving eigenvalue problems are introduced and, as a consequence of their scale invariance, eigenvectors are not isolated in vector spaces and each eigendirection defines a corresponding vector subspace. In another dimension of the problem, numerical computations ask for solution sets consisting only of isolated points in the state space, and an adequate solution may be the imposition of norm equality constraints on iterates of the method. Accordingly, the resulting spherical search space is an embedded submanifold of the original linear space. It is easy to perceive that working inside search spaces that carry the structure of manifolds may create certain difficulties in algorithm construction. Very often, iterative optimization algorithms are strongly dependent on the linear space structure of a given search space because a new iterate is generated by adding update increments to previous iterates in order to reduce the values of the cost function. The update direction and step size are frequently computed using a local model of the cost function, often based on first and second derivatives of that function. To establish similar algorithms on manifolds, the respective operations must be translated, at least, into the language of Topology—results presented so far need even more structure, using concepts of Riemannian geometry to construct the proper extensions. Furthermore, expanding optimization algorithms to abstract manifolds is only the first step toward the final target, because finding efficient numerical procedures able to implement them is a second step, validating the first part of the effort.

3.2 Basic Aspects of Manifolds

Manifolds are, in general terms, spaces that locally look like Euclidean spaces $\mathbb{R}^n$, and on which it is possible to define calculus-like operations, like differentiation and integration. Among the most familiar examples, apart from Euclidean spaces themselves, are straight lines, circles, spheres, ellipsoids, and cylinders. In higher dimensions, we have examples such as the n-sphere $\mathbb{S}^n$ and graphs of differentiable maps between Euclidean spaces. The most fundamental kind of manifold can be considered the topological one, that is defined as a topological space possessing certain properties conveying what we mean when it is said that it locally looks like $\mathbb{R}^n$. In addition, a large number of applications of manifolds involve concepts that could be considered extensions of operators found in widely known multivariable calculus. This occurs when applying manifold theory to geometry and studying properties as volume and curvature. Volumes are usually computed by means of some kind of integration, and curvatures are calculated with formulas involving second derivatives—for this reason extending these ideas to manifolds would require some means of defining differentiation and integration on them. As further examples, applications of manifold theory to classical mechanics demand solving systems of ordinary differential equations on configuration spaces that can be modeled by manifolds, and the application to general relativity involves solving systems of partial differential equations, evidencing again the need for calculus. The most essential requirement for transferring the ideas of calculus to manifolds is some notion of differentiability. In Euclidean spaces, it is relatively easy to express that concept on an intuitive level as, for example, we can call a curve differentiable (or smooth) if it has a tangent line that varies continuously from point to point, and a smooth surface could be one that has a tangent plane that varies smoothly. More complex applications, however, may require that certain manifolds not be subsets of Euclidean spaces. It is a fact that it would be a big benefit to be able to work with manifolds viewed as topological spaces, without restrictions to ambient spaces like $\mathbb{R}^n$. So, there is a need to think of manifolds as autonomous spaces, not necessarily subsets of larger spaces. As can be seen from the definition below, purely topological properties are not sufficient to define the desired smoothness profile, and topological manifolds will not be enough for that purpose (extending calculus on $\mathbb{R}^n$). Therefore, it is needed to define smooth manifolds, which are objects having two layers of structure—topological and differential.

3.2.1 Topological Manifolds

A topological manifold of dimension n is a Hausdorff topological space that is second countable and locally Euclidean of dimension n, that is, every point has a neighborhood that is homeomorphic to an open subset of $\mathbb{R}^n$. A good reason for imposing these properties is that manifolds tend to behave in ways more similar to common experience with Euclidean spaces. A simple example is $\mathbb{R}^n$ itself, as it is easy to see—as a metric space, it is Hausdorff too; taking the set of all open balls with rational centers and rational radii as a countable basis, it is possible to show that it is second countable; and, of course, any point $p \in \mathbb{R}^n$ belongs to an open ball that is homeomorphic to itself.

Let M be a topological manifold of dimension n. A coordinate chart on M is a pair (U, φ), where U is an open subset of M and $\varphi : U \rightarrow U'$ is a homeomorphism from U to an open subset $U' = \varphi(U) \subset \mathbb{R}^n$, as illustrated in Fig. 3.1. According to the definition, each point $p \in M$ belongs to the domain of some chart (U_p, φ_p).

So, lines and curves are examples of 1-dimensional manifolds—the real line may be considered the simplest example. Other examples are given by plane curves such as circles, straight lines, parabolas, or the graphs of continuous functions. Smooth space curves, which are often described in parametric format, offer another class of examples. In each case, a point in the manifold can be unambiguously determined by a single real number. So, for example, a point on the real line is represented by a real number, and it is possible to identify a point on the circle by an angle, a point on a graph by its x-coordinate, and a point on a parametric curve by the corresponding parameter.

Although a parameter value determines a point, it (the point) may be associated to different parameter values. However, within appropriate neighborhoods, there exists a one-to-one correspondence between nearby real numbers and associated points on lines or curves. Following this line, 2-dimensional manifolds are surfaces, and common examples are planes, spheres, cylinders, ellipsoids, paraboloids, hyperboloids.

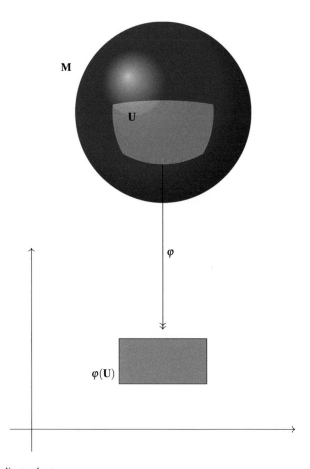

Fig. 3.1 Coordinate chart

In these manifolds, two coordinates are necessary to completely identify a point. Hence, in a plane we can use Cartesian or polar coordinates, and for spheres latitude and longitude are sufficient. As in 1-dimensional manifolds, the correspondence between points and pairs of numbers is usually local.

3.3 Proposed Method

Taking into account that manifolds are more general environments than Euclidean spaces, and most global optimization algorithms are designed to deal with problems defined on the latter ones, it is reasonable to try to enlarge the reach of current optimization methods, given the wealth and complexity of problems faced by researchers nowadays. This task already begun [1, 7, 8] helped by tools of Differential Geometry and Topology, and in [1] it is focused at the optimization of differentiable functions on certain types of smooth manifolds—previously published work is really impressive and the application of well-established mathematical results on manifold theory is effected in a very ingenious way, giving rise to useful optimization algorithms.

In this chapter it is described a more general idea, directed toward optimizing not necessarily differentiable functions defined on topological manifolds by using metaheuristic methods whose candidate populations originally evolve in Euclidean configuration spaces. This framework allows previously tested methods to extend their reach by making small adaptations.

3.3.1 Statement of the Problem

The objective is to find a global minimizer $\mathbf{x}^* \in \mathbf{M}$ for a given objective function $f : \mathbf{M} \to \mathbb{R}$, where $\mathbf{M}$ is a finite dimensional topological n-manifold, covered by a finite number (N_c) of coordinate domains $\{U_i : i = 1, \ldots, N_c\}$, and associated with a finite number of coordinate charts $\{(U_i, \varphi_i) : i = 1, \ldots, N_c\}$.

An extra condition is that the images of the coordinate domains $\varphi_i(U_i)$ be open hyper-rectangles of $\mathbb{R}^n$. This hypothesis does not represent a true limitation, considering that in most relevant settings we have manifolds in which the $\varphi_i(U_i)$ are homeomorphic, or even diffeomorphic, to open hyper-rectangles. This important assumption will enable algorithms that originally evolve their populations in the interior of hyper-rectangles to be applied without significant alteration, so that anterior accumulated knowledge will not be lost. Figure 3.2 illustrates the described configuration.

When applying the proposed method, one preliminary measure is necessary. Given that all images $\varphi_i(U_i) \subset \mathbb{R}^n$ are hyper-rectangles, it is simple to substitute them, without loss of generality, for only one set, say, $\mathbf{H} \overset{\triangle}{=} (a_1, b_1) \times (a_2, b_2) \times \cdots \times (a_n, b_n)$, as open hyper-rectangles are diffeomorphic among themselves when $\mathbb{R}^n$ is

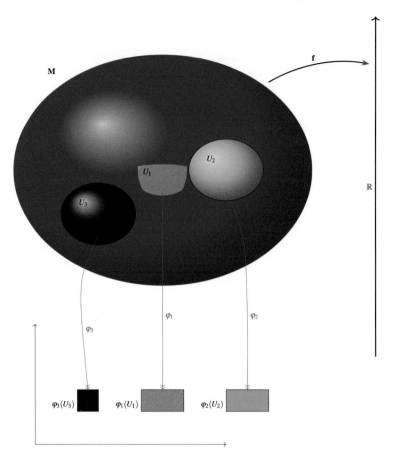

Fig. 3.2 Setting for the global optimization problem

endowed with its standard topological and differential structures. As expected, the corresponding identifying maps (diffeomorphisms/homeomorphisms between the $\varphi_i(U_i)$ and $(a_1, b_1) \times (a_2, b_2) \times \cdots \times (a_n, b_n)$) must be composed with the original charts in order to not distort the final quantitative results.

3.3.2 Proposal for Global Optimization on Manifolds

At this point it is possible to state a global minimization method on **M**, assuming that a metaheuristic algorithm capable of globally minimizing functions defined on hyper-rectangles of $\mathbb{R}^n$ is available. Figure 3.3 illustrates the scheme.

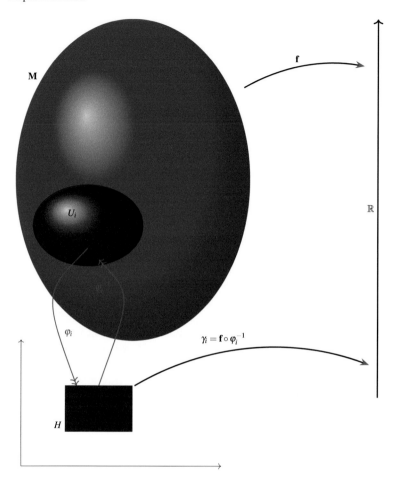

Fig. 3.3 Basic composite functions

- **Preparation**

  - Find the analytical expressions for the inverses φ_i^{-1} of coordinate maps φ_i ($i = 1, \ldots, N_c$);
  - Find the analytical expressions for the composite functions $\gamma_i = f \circ \varphi_i^{-1} : \varphi(U_i) = \mathbf{H} \rightarrow \mathbb{R}$ ($i = 1, \ldots, N_c$), where $f : \mathbf{M} \rightarrow \mathbb{R}$ is the original cost function and $\mathbf{H} = (a_1, b_1) \times (a_2, b_2) \times \cdots \times (a_n, b_n)$, as above;
  - Define a new cost function by $\Gamma \overset{\Delta}{=} min\{\gamma_i : i = 1, \ldots, N_c\}$.

- **Step 1—generation of new points according to the driving evolutionary technique**

  - Execute a new iteration, generating a new population or single candidate point, driven by Γ values.

- **Step 2—decision making**

  – If convergence criteria are met or number of iterations is over, go to Final step,
    else go to Step 1.

- **Final step**

  – Calculate $\varphi_i^{-1}(\mathbf{x}_0^*)$, where $\mathbf{x}_0^*$ is the final result of previous steps, expected to be
    one of the global minimizers of Γ. In consequence, $\varphi_i^{-1}(\mathbf{x}_0^*)$ is assumed to be
    one of the minimizers for the original f.

Due to the high flexibility of the overall algorithm, and depending on the driving
global optimization method, it is possible to change Γ.

Considering the architecture of the method, it becomes feasible to visit all regions
of $\mathbf{M}$ (in parallel, if necessary) and evolve candidate populations directly inside it,
without using equality constraints or similar devices—when points in $\mathbf{H}$ evolve, they
give rise to points inside $\mathbf{M}$.

When implementing the method, it is possible, for example, to launch N_c program
threads aiming at finding minimizers in each coordinate domain and, at the end,
choose the best one; however, if the number of charts is too big such a procedure
might not be that efficient, but it usually is a feasible alternative. Furthermore, such
an approach opens the way to eliminate equality constraints in constrained global
optimization problems, simultaneously reducing the dimension of the search space.
As said above, the paradigm will be demonstrated by means of the Fuzzy ASA
method [4], to be briefly described in the next section. After that, some numerical
simulations will help in assessing the quality of the approach.

3.4 Fuzzy Adaptive Simulated Annealing (Fuzzy ASA)

The Fuzzy ASA algorithm [4] may be viewed as a fuzzy-controlled version of the
ASA method [9] that, by its turn, is inspired by the simulated annealing concept,
but featuring a significant number of improvements. In this section some of its most
important aspects will be highlighted.

3.4.1 Re-annealing

It is the ability to dynamically re-scale the parametric temperatures, adapting generating probability distribution functions relative to each dimension according to
different sensitivities. In simple terms, if the cost function does not show significant
variations whenever a given parameter is altered, it may be sensible to extend the
search interval amplitude in that dimension in particular, and vice-versa.

3.4.2 Quenching

The standard ASA design offers the possibility of adjusting several parameters related to the quenching process, allowing the user or any automatic control mechanism to change the default behavior of the "cooling" process that drives the evolution of parametric temperatures. In practice, this may be useful in case of stagnation near suboptimal regions.

3.4.3 High Structural Flexibility

The ASA system was constructed so as to allow users to alter many subsystems with little or zero programming effort. Therefore, it is possible to change the behavior of generation/acceptance mechanisms, quenching degree, termination criteria, seed generation, for example, by changing certain numerical control variables and/or redirecting mainstream execution flow to customized routines.

3.4.4 More Relevant Characteristics

Although the default search space is a hyper-rectangle, its compactness is not a limitation and, even without previous information about global optima location, it is enough to choose sufficiently ample hyper-rectangular domains.

Furthermore, the quenching mechanism is often able to improve the speed of convergence, although there is the possibility of reaching prematurely non-global extrema. In some settings, however, it might not be possible to find alternative ways out of stagnation occurrence, as sometimes happens to functions operating in high-dimensional spaces. To cope with this problem, a fuzzy controller was idealized (Fuzzy ASA) [2, 3]. In this approach the standard ASA system is faced as a MISO

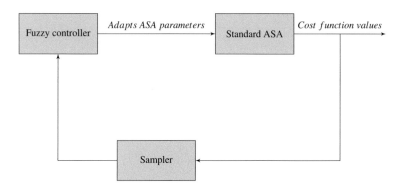

Fig. 3.4 The overall structure of Fuzzy ASA

(Multiple Input Single Output) dynamical system, and the new code simply implements a closed-loop feedback system by sampling its output (current value of objective function) and acting on its inputs (a subset of run-time adjustable parameters, related to the quenching process) according to a fuzzy inference system, mimicking human behavior whenever subject to a similar context. Hence, run-time control can make temperature evolution slower or faster, in addition to being able to take corrective actions in case of premature convergence. Fuzzy ASA code raises the quenching degree after detecting decreasing optimization performance or stagnation trend, striving to recover from a possible convergence to non-global optima. Figure 3.4 depicts a diagram portraying Fuzzy ASA's working principle.

3.4.5 Additional Comments About Internal Structure

It is well-known that standard simulated annealing implementations may present low convergence speed, being their average performance not very encouraging [6, 10, 14, 15]. However, there are ways to overcome the shortcomings of most annealing schemes, such as ASA, that is a sophisticated and really effective global optimization method [9]. The ASA method is a very powerful tool for applications involving difficult objective functions, as those related to neuro-fuzzy systems design and optimization-based design, for example, considering its performance and simplicity. Having the benefits of being publicly available, adjustable and well-maintained, it is an alternative to other global optimization tools, according to the published benchmarks, that demonstrate its good quality. On the negative side, stochastic global optimization algorithms share some inconvenient features, like spending large periods of time with poor improvement when approaching global extrema. In simulated annealing implementations, this may be attributed to the cooling schedule, whose speed is limited by the characteristics of probability density functions used to generate additional candidate points. Hence, if we choose to employ, say, Boltzmann annealing, the temperature has to be lowered at a maximum rate of $T(k) = T(0)/ln(k)$. In the fast annealing paradigm, the schedule changes to $T(k) = T(0)/k$, if assurance of convergence with probability 1 is to be kept, resulting in a faster schedule. ASA's scheme is even better, and given by

$$T_i(k) = T_i(0)exp(-C_i k^{1/D}) \qquad (3.1)$$

where C_i is a user-defined constant, due to its improved generating distribution. Note that the subscripts indicate independent evolution of temperatures for each dimension. As cited above, it is also possible for the ASA user to take advantage of simulated quenching, by means of the expression

$$T_i(k) = T_i(0)exp(-C_i k^{Q_i/D}) \qquad (3.2)$$

where Q_i is the quenching parameter corresponding to dimension i.

By choosing quenching parameters with values greater than one, higher evolution speeds are possible, but the convergence to a global optimum is no longer assured (please, see [9]). Such a procedure is especially recommended for high-dimensional parameter spaces.

Simulated annealing algorithms use three fundamental components, that have large impact on the final implementation:

- A probability density function g(.), used in the generation of new candidate points;
- A PDF a(.), used in the acceptance/rejection of new generated points;
- A schedule T(.), that determines how the temperatures will vary during the execution of the algorithm.

The usual approach is to choose a starting point and set the initial temperature so that the state space may be sufficiently explored. Following this, new points are iteratively computed according to the generating PDF g (.) and probabilistically accepted or rejected, by using the PDF a(.). In case of acceptance, they substitute current base points. During the execution, temperatures are reduced and this lowers the probability of acceptance of new generated points with higher cost values than those corresponding to current points. However, there is still a positive probability of accepting such points, opening the possibility of escaping from local minima.

Having been designed to find global minima inside a compact subset of n-dimensional Euclidean space, ASA generates points componentwise, as shown below

$$x_{i+1} = x_i + \Delta x_i, \quad with \quad \Delta x_i = y_i(B_i - A_i),$$

$[A_i, B_i]$ = interval corresponding to ith dimension,
$y_i \in [-1, 1]$ is given by $y_i = sign(u_i - 1/2)T_i[(1 + 1/T_i)^{|2u_i - 1|} - 1]$,
where $u_i \in [0, 1]$ is generated by means of the uniform distribution, and
T_i = current temperature relative to dimension i.

At this point, it is possible to identify two main issues related to Fuzzy ASA design: the controller must be able to use actual cost function values in order to infer the current status of the optimization in progress and needs to alter ASA parameters so as to overcome inconvenient situations, like permanence near non-global minima or slow progress.

The former point may be handled by using the concept of sub-energy function, used in the TRUST method [5].

It is given by the expression

$$S(x, x_0) = log\ (1/[1 + exp(-(f(x) - f(x_0)) - a)]) \tag{3.3}$$

where $a \in \mathbb{R}$ and x_0 is the current base point.

The base point is defined as the one corresponding to the minimum value found so far in the progressing simulation. Hence, function S behaves like the original f when the search evolves toward regions under the current minimum, and tends to be flat in upper ones. In this way it is possible to obtain the relative position between current and base points by inspecting values obtained from the sub-energy function. Such a scheme allows us to synthesize approximate conclusions as *"The search is NEAR the current minimum"* or *"The search is VERY FAR from the current minimum"*, giving rise to a possibility of fuzzy modeling.

The second problem is related to the consequent stages of the fuzzy rule base, in which it is necessary to provide corrective actions in order to keep the search toward the global minimum. This measure was taken by varying quenching degrees for generating and acceptance probability distribution functions. To reach that aim, individual quenching factors for each dimension and one cost quenching factor were employed.

3.5 Computational Evaluation

To assess the efficacy of the presented method, three global optimization simulations with difficult objective functions and simple manifolds are used. By doing so, it is simultaneously possible to measure the optimization power of the algorithm and illustrate some implementation details. For the sake of comparison, in each case the results corresponding to the proposed paradigm and the traditional one (using penalty functions) are shown, the latter imposing penalties associated to equality constraints, in order to keep evolving points inside the domain manifold. Please, note that in the former method the primary generation of candidate points will take place in a region contained in a Euclidean space whose dimension is smaller than that in the latter. So, the comparison will be based on the number of objective function evaluations necessary to approximate the global minimizer.

3.5.1 Auxiliary Manifolds Used in the Tests

Despite the fact that the presented techniques may be used with any topological manifold, we chose homeomorphic copies of $\mathbb{S}^n$, the unitary hypersphere contained in $\mathbb{R}^{n+1}$—the charts composing the corresponding atlas and that appear in subsequent experiments are described below. Although there are other atlases that make $\mathbb{S}^n$ into a topological manifold, the chosen one certainly is a good choice to the intended task. In this case $\mathbb{S}^n$ receives $2 \times (n+1)$ charts $\{(U_i^+, \varphi_i^+) : i = 1, \ldots, n+1\}$ and $\{(U_i^-, \varphi_i^-) : i = 1, \ldots, n+1\}$, established in [11] by the following expressions

$$U_i^+ \triangleq \{(x_1, x_2, \ldots, x_n, x_{n+1}) \in \mathbb{S}^n : x_i > 0\} \tag{3.4}$$

$$U_i^- \triangleq \{(x_1, x_2, \ldots, x_n, x_{n+1}) \in \mathbb{S}^n : x_i < 0\} \tag{3.5}$$

$$\varphi_i^-(x_1, \ldots, x_{n+1}) = \varphi_i^+(x_1, \ldots, x_{n+1}) \overset{\Delta}{=} (x_1, \ldots, x_{i-1}, x_{i+1} \ldots, x_{n+1}) \tag{3.6}$$

where $\varphi_i^- : U_i^- \rightarrow \mathbb{R}^n$ and $\varphi_i^+ : U_i^+ \rightarrow \mathbb{R}^n$.

The inverse mappings are given by

$$(\varphi_i^-)^{-1}(v_1, \ldots, v_n) = \left(v_1, \ldots, v_{i-1}, -\sqrt{1 - |v|^2}, v_i \ldots, v_n\right) \tag{3.7}$$

$$(\varphi_i^+)^{-1}(v_1, \ldots, v_n) = \left(v_1, \ldots, v_{i-1}, \sqrt{1 - |v|^2}, v_i \ldots, v_n\right) \tag{3.8}$$

3.5.2 Example—Ackley Function Constrained to 2-Dimensional Sphere and Viewed as a Submanifold of $\mathbb{R}^3$

In this example the intention is to minimize the Ackley function (Fig. 3.5), whose expression is given by

$$f(\mathbf{x}) \overset{\Delta}{=} -20 exp\left(-0.2\sqrt{\frac{\sum_{i=1}^n x_i^2}{n}}\right) - exp\left(\frac{\sum_{i=1}^n cos(2\pi x_i)}{n}\right) + 20 + e \tag{3.9}$$

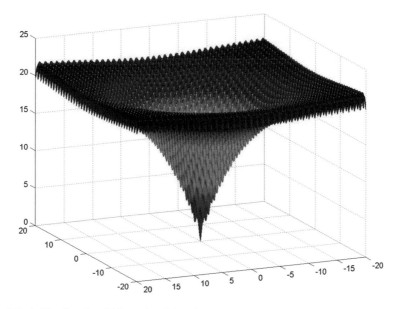

Fig. 3.5 Ackley function (bidimensional domain)

where n is the domain dimension, $\mathbf{x} = (x_1, x_2, \ldots, x_n) \in \mathbb{R}^n$ and $-32.768 \leq x_i \leq 32.768$, $i \in \{1, \ldots, n\}$ restricted to the surface of a 2-dimensional sphere which contains the global minimizer in $\mathbb{R}^3$ itself, making it easier to evaluate whether the algorithm is able to find the global minimizer. The constraining sphere has radius 13 and center at $(0,0,13)$.

This function has a global minimizer (in the above specified domain) at $\mathbf{x}^* = (0, 0, 0)$ with value 0.

In this experiment we chose to work in $\mathbb{R}^3$, while the proposed algorithm will make ASA evolve in a 2-dimensional region, taking into account the intrinsic dimensional reduction (equal to the codimension of $\mathbb{S}^2$ in $\mathbb{R}^3$) made possible by the 2-dimensional charts. After 50 executions of each type of test, the presented method converged to the global minimizer in 100 % of the cases, and the penalized problem in only 50 % (25 runs). In all tests the performance was better when using the standard ASA

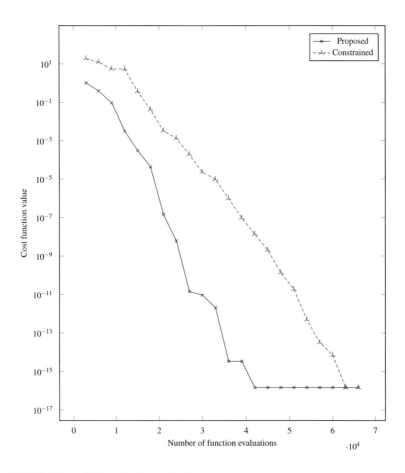

Fig. 3.6 Minimizing Ackley function on 2-spheres

implementation. Below, it is shown a comparative graph displaying the evolution of typical runs for each type of method.

By examining Fig. 3.6, it is easy to see that both types converge to the global minimum, but the presented method spends less function evaluations to reach the target. Naturally, there are different effective cost functions in each type of experiment, considering that in the "standard" penalized optimization procedure it is necessary to include the calculation of the constraints themselves, and in the presented method there are additional function evaluations, relative to the charts covering the underlying manifold. In practice the difference in terms of unitary computational effort is not significant. In any case, there is a big benefit in evolving directly "inside" the natural domain of a given problem, not having to deal with equality constraints and associated penalty functions, for instance.

3.5.3 Example—Griewank Function Restricted to a 2-Dimensional Sphere, Viewed as a Submanifold of $\mathbb{R}^3$

The current task is to minimize the Griewank function, defined by

$$f(\mathbf{x}) \overset{\Delta}{=} \frac{1}{4000} \sum_{i=1}^{n} x_i^2 - \prod_{i=1}^{n} cos\left(\frac{x_i}{\sqrt{i}}\right) + 1 \qquad (3.10)$$

where n is the domain dimension, $\mathbf{x} = (x_1, x_2, \ldots, x_n) \in \mathbb{R}^n$ and $-600 \leq x_i \leq 600$, $i \in \{1, \ldots, n\}$.

It is constrained to the surface of a 2-dimensional sphere containing the global minimizer in $\mathbb{R}^3$. As in the previous example, the idea is to make it feasible for the algorithm to find the desired state. Once again, the sphere has radius 13 and center at (0,0,13).

This function has a global minimizer (in the specified domain) at $\mathbf{x}^* = (0, 0, 0)$ with value 0.

Although $n = 3$ in this test, the algorithm makes ASA evolve in a 2-dimensional environment, considering the dimensional reduction (equal to the codimension of $\mathbb{S}^2$ in $\mathbb{R}^3$) that occurs thanks to the 2-dimensional charts. After 50 executions of each type of test, the presented method converges to the global minimizer in 100 % of the cases, and the penalized problem in only 2 %. In both cases, overall performance was better when using the original ASA implementation. Below, the comparative graph (Fig. 3.7) shows the evolution of the best run of the constrained type and a typical one relative to the proposed method.

Figure 3.7 demonstrates that both methods arrive at the desired minimum, however the proposed method takes less function evaluations to converge. As stated before, there are different effective cost functions in each type of execution, as in the conventional constrained optimization it is necessary to incorporate the calculation

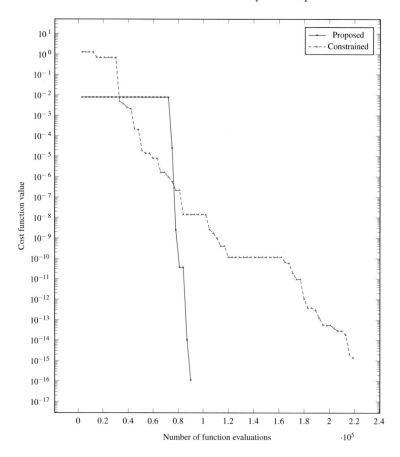

Fig. 3.7 Minimizing Griewank function on 2-spheres

of the constraints, and in the presented method there are extra original cost function
evaluations, corresponding to the several charts covering the manifold. Anyway, the
difference in terms of unitary computational effort is not significant.

3.5.4 Example—Griewank Function Restricted to a 3-Dimensional Sphere, Considered a Submanifold of $\mathbb{R}^4$

This experiment is aimed at minimizing the Griewank function (Fig. 3.8), defined as above, but now the domain will be restricted to the surface of a 3-dimensional sphere containing the global minimizer in $\mathbb{R}^4$. Here $n = 4$, and the sphere has radius 10 with the center located at $(0,0,0,10)$. The global minimizer (inside the established domain) is again the origin $\mathbf{x}^* = (0, 0, 0, 0)$ with value 0.

Once more the proposed algorithm drives Fuzzy ASA so as to evolve in a 3-dimensional region, considering the intrinsic dimensional reduction (equal to the codimension of $\mathbb{S}^3$ in $\mathbb{R}^4$) given by the 3-dimensional charts. After 50 executions of each type of test, the proposed method converged to the global minimizer in 90 % of the cases, and the penalized problem only approached the global minimizer in 2 % (1 execution instance). In both cases, overall performance was better when using the Fuzzy ASA implementation. Below, a comparative graph (Fig. 3.9) displays the evolution of the best run of the constrained type and a typical one relative to the presented method.

Figure 3.9 shows that only the present method really arrived at the desired minimum, spending less function evaluations to get to a convergence state. On the other hand, the hit rate of the classical constrained method was not satisfactory because after all experiments it just got nearby the true minimizer and did not actually con-

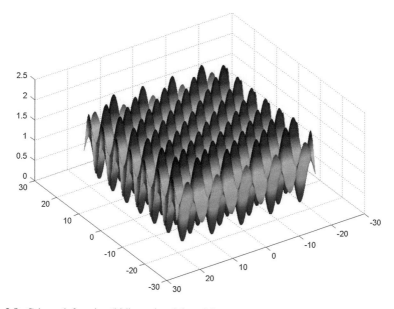

Fig. 3.8 Griewank function (bidimensional domain)

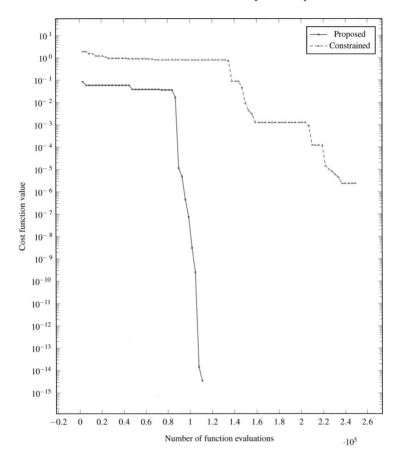

Fig. 3.9 Minimizing Griewank function on a 3-dimensional sphere

verge to it. As before, we have different effective cost functions in each type of execution, considering that in the conventional constrained optimization it is necessary to incorporate the calculation of the constraints themselves, and in the presented method there are extra original cost function evaluations, corresponding to the several charts associated to the underlying manifold.

3.6 Conclusion

This chapter presented an alternative approach to evolutionary global optimization on topological manifolds that allows to process functions whose domains are not linear spaces—the evolution of populations or candidate points takes place directly in those regions. A significant aspect is that many already existing evolutionary methods can take advantage of the proposed method without changes in their original structure. The range of applications is enlarged considering that, apart from the original purpose, it is possible to achieve dimensional reduction in constrained optimization. In addition, according to the presented results it is possible to infer that the method is effective and can avoid problems present in techniques aiming at the same target as, for example, adding penalizing terms in order to force evolving populations to stay inside some specific region of the state space (penalty functions). When compared to existing techniques aimed at optimization on manifolds [1], the method shows more generality because it is capable of handling problems defined on topological manifolds—they do not need to be smooth.

Furthermore, previously existing algorithms are usually complex, while the present one is simpler in the sense that it is able to deal with nonsmoothness in a natural way, considering the characteristics inherent in metaheuristics. Furthermore, the proposed scope is larger in a different way, noting that the algorithm is capable of dealing with nondifferentiable functions.

References

1. Absil, P.-A., Mahony, R., Sepulchre, R.: Optimization Algorithms on Matrix Manifolds. Princeton University Press, Princeton (2008)
2. Aguiar e Oliveira Jr, H., Petraglia, A.: Global Optimization Using Space-Filling Curves and Measure-Preserving Transformations. In: Gaspar-Cunha, A., et al. (eds.) Soft Computing in Industrial Applications, AISC 96, pp. 121–130. Springer, Berlin (2011)
3. Aguiar e Oliveira Jr, H., Petraglia, A.: Global optimization using dimensional jumping and fuzzy adaptive simulated annealing. Appl. Soft Comput. **11**, 4175–4182 (2011)
4. Aguiar e Oliveira Jr, H., Ingber, L., Petraglia, A., Petraglia, M.R., Machado, M.A.S.: Stochastic Global Optimization and Its Applications with Fuzzy Adaptive Simulated Annealing. Springer, Berlin (2012)
5. Barhen, J., Protopopescu, V., Reister, D.: TRUST: a deterministic algorithm for global optimization. Science **276**, 1094–1097 (1997)
6. Corana, A., Marchesi, M., Martini, C., Ridella, S.: Minimizing multimodal functions of continuous variables with the simulated annealing algorithm. ACM Trans. Math. Softw. **13**, 262–280 (1987)
7. Helmke, U., Moore, J.B.: Optimization and Dynamical Systems. Springer, London (1994)
8. Hillermeier, C.: Nonlinear Multiobjective Optimization—A Generalized Homotopy Approach. Birkhäuser, Basel (2001)
9. Ingber, L.: Adaptive simulated annealing (ASA): lessons learned. Control Cybern. **25**(1), 33–54 (1996)
10. Kirkpatrick, S., Gelatt Jr, C.D., Vecchi, M.P.: Optimization by simulated annealing. Science **220**, 671–680 (1983)

11. Lee, J.M.: Introduction to Smooth Manifolds, 2nd edn. Springer, New York (2013)
12. Ma, Y., Fu, Y.: Manifold Learning Theory and Applications. CRC Press, Boca Raton (2012)
13. Nocedal, J., Wright, S.J.: Numerical Optimization. Springer, New York (1999)
14. van Laarhoven, P.J.M., Aarts, E.H.L.: Simulated Annealing: Theory and Applications. D. Rei-
    del, Dordrecht (1987)
15. Weise, T.: Global Optimization Algorithms—Theory and Application. http://www.it-weise.
    de/. Accessed 11 Mar 2015

Chapter 4
Constrained Global Optimization on Manifolds

Abstract This chapter presents techniques for dealing with constrained global optimization of real valued functions defined on smooth manifolds, subject to equality constraints. Functional constraints must satisfy certain smoothness conditions, not excluding simultaneous restrictions of different types, being the effect of dimensional reduction proportional to the number of equality restrictions. The problems under study do not need to restrict cost functions to be differentiable or even continuous, and the optimization task is done so as to keep candidate points inside corresponding submanifolds, evolving therein along the optimization process. The techniques may be employed together with an extensive family of already tested evolutionary methods and, after introducing the fundamental ideas, selected examples will demonstrate the effectiveness of the presented methods.

4.1 Introduction

Although constrained global optimization on linear spaces is a very developed research area, with many techniques able to address a multitude of problems in this field, in this chapter it is presented an alternative viewpoint, capable of facing the problem in a efficient way, in the sense that dimensional reduction is feasible, provided that some theoretical conditions be true. To get started, we assume that there is a constrained optimization problem involving real valued functions defined on manifolds and subject to equality and inequality constraints, and it is to be solved by means of an evolutionary method. We will see that equality constraints give origin to vector valued functions which can define submanifolds of the original domain, inside which such constraints are automatically satisfied. Hence, by generating candidate points directly into the lower dimensional submanifold, we will provoke an apparent elimination of equality constraints and significant reduction in terms of computational effort, considering that the evolution of candidate points will happen in a lower dimensional context. So, provided that the necessary conditions are satisfied, the reduction in the number of dimensions equals the number of equality constraints. In the recent past, important research efforts related to optimization on manifolds [2, 6] were done and they provided alternatives to a few general purpose

© Springer International Publishing Switzerland 2016
H. Aguiar e Oliveira Junior, *Evolutionary Global Optimization,*
Manifolds and Applications, Studies in Systems, Decision and Control 43,
DOI 10.1007/978-3-319-26467-7_4

constrained optimization algorithms. Techniques capable of working on manifolds may feature less computational complexity, and frequently show better numerical properties, being capable of avoiding suboptimal attraction regions. At this point, an approach for constrained global optimization of functions defined on finite dimensional manifolds with equality constraints will be presented. The equations defining equality restrictions allow us to obtain the desired original domain simplification, as will be explained below. It is possible to think of manifolds as spaces that are locally similar to linear spaces. Following this reasoning, $\mathbb{R}^n$ is a manifold as well. To proceed, however, it will be necessary to present some topics of General and Differential Topology that will help in the understanding of the proposed optimization framework. After the preparatory part, it will be possible to see that most existing evolutionary global optimization heuristics may be combined with the paradigm. Having in mind that many important problems may be viewed as models whose defining parameters evolve on submanifolds of certain spaces, further initiatives in that direction are surely useful. The method can be applied to hard optimization problems, defined on regions modeled as smooth manifolds. The smoothness imposition on manifolds and defining equations is important because the intention is to use auxiliary results that depend on such premises. At present it is possible to find results whose scopes are somewhat restricted, many of them related to matrix optimization problems. The underlying themes are relatively common in areas related to Linear Algebra, signal processing, data mining, and statistical analysis [15], frequently using particular characteristics of each context, for the sake of improving final results. It is also worth to mention that the application of evolutionary methods to this field is still in its very beginning, what gives us one more reason to investigate new possibilities of solution. It is expected that the work on manifolds might create certain difficulties in numerical algorithm synthesis. This is so mainly because iterative optimization algorithms depend on the vector space structure of cost functions' domains, and new candidate points are generated by adding update increments to previous iterates so as to reduce functions values after each step. Another typical aspect is that the search direction and step size are usually calculated by means of a local model of the cost function, often based on derivatives. When dealing with smooth manifolds these operations must be extended beyond the realm of vector spaces, to the language of Differential Topology. This transformation is not that simple and needs significant research efforts. In addition, the theoretical extension of optimization algorithms to the context of manifolds is just the initial stage, considering that the existence of an efficient numerical procedure is an essential factor for the solution of the original problem.

Currently it is possible to find solid optimization algorithms resulting from the combination of well-known results of Topology and Differential Geometry. For example, there are several methods designed by means of the concepts of retraction and vector transport, which are relaxations of the geometric concepts of motion along geodesics and parallel transport [1, 2, 4, 7, 11].

4.2 Basic Characteristics of Smooth Manifolds

In general terms, smooth manifolds are objects that locally look like Euclidean spaces $\mathbb{R}^n$, and on which it is possible to define operations that extend those ones used in classic Calculus. The most familiar examples are $\mathbb{R}$, circles, spheres, and ellipsoids. In addition, there are examples like the n-sphere $\mathbb{S}^n$ and graphs of smooth maps between linear spaces.

As expected, several significant applications of manifolds involve operations similar to the ones found in differential calculus, as the use of manifold theory in differential geometry, that investigates properties of concepts like volume and curvature. In the same way as in $\mathbb{R}^n$, volumes are usually computed by means of some kind of integration, and curvatures are calculated with expressions including operations similar to second derivatives, for instance. Hence, the generalization of such ideas to manifolds requires that, somehow, a way to define differentiation and integration operations on manifolds be found. As further examples, applications to classical mechanics involve solving systems of ordinary differential equations on spaces modeled by means of manifolds, and the study of general relativity involves the solution of systems of partial differential equations—the need for calculus shows up again. So, the first barrier to overcome is to establish a formal definition for such concepts in the realm of manifolds. In $\mathbb{R}^n$ it is not difficult to describe the meaning of smoothness. For example, it is possible to call a curve smooth if each point has a tangent line varying continuously. In the same way, a smooth surface would have tangent planes moving smoothly along its extension. In many settings, however, it is necessary not to restrict the working environment to subsets of Euclidean spaces. Therefore, in these cases it is necessary to be able to work with manifolds as topological spaces, without being restricted to $\mathbb{R}^n$. This is so because in certain applications there are models in which no physical meaning can be associated to higher dimensional spaces containing certain submanifolds. As cited previously, purely topological properties are not enough to define the desired smoothness properties and, consequently, topological manifolds will not be enough for that purpose (extending Calculus on $\mathbb{R}^n$). Accordingly, smooth manifolds are objects that have two layers of structure—topological and differential.

4.2.1 Smooth Manifolds

Since the definition of (topological) manifold given above is only sufficient for studying topological properties, there are no adequate devices for making quantitative operations or similar things. To add the necessary elements that will provide the technical basis for the definition of derivatives of functions, curves, or maps defined on manifolds, it is essential to introduce a new kind of manifold—a smooth (or C^∞) manifold. Unfortunately, it is not sufficient to define a smooth manifold simply as a topological manifold with special attributes because smoothness is not always invariant under homeomorphisms. Hence, a sound definition for the concept of smooth manifold should include extra structures beyond its topology, making it possible to decide which functions defined on it are differentiable, for instance. To discover what this additional structure may be, let M be a topological manifold of dimension n. It is assumed that each point in M belongs to the domain of a coordinate map $\varphi : U \to U' = \varphi(U) \subset \mathbb{R}^n$. In order to formulate the definition of smoooth manifolds, certain concepts related to Euclidean spaces are needed, so let us remember some basic definitions.

Let U and V be open subsets of $\mathbb{R}^n$ and $\mathbb{R}^m$, respectively; a map $F : U \to V$ is smooth if each of the component functions of F has continuous partial derivatives of all orders. If F is bijective and has a smooth inverse, it is named a diffeomorphism, which is a homeomorphism as well.

Note that if M is a topological manifold and (U, φ_U) and (V, φ_V) are two charts such that $U \cap V \neq \emptyset$, then the composite map $\varphi_V \circ \varphi_U^{-1}$ (the transition map) is a composition of homeomorphisms, and so is a homeomorphism too.

Definition 4.1 Two charts (U, ϕ) and (V, ξ) are *smoothly compatible* if either $U \cap V = \emptyset$ or the transition map $\xi \circ \phi^{-1} : \phi(U \cap V) \to \xi(U \cap V)$ is a diffeomorphism (see Fig. 4.1). An atlas for M is a collection of charts whose domains cover M, and it is said to be smooth if any two charts in it are smoothly compatible with each other. A smooth atlas on M is maximal if it is not contained in a larger smooth atlas. So, any chart that is smoothly compatible with every chart in it is already inside. A smooth structure on a topological n-manifold M is a maximal smooth atlas.

Definition 4.2 A *smooth manifold* is a pair (M,A), where M is a topological manifold and A a smooth structure on M. It is common practice to refer to the smooth manifold as M.

In this fashion, the term **smooth manifold structure** indicates a manifold topology coupled to a smooth structure. It is important to highlight that smooth structures are extra objects that must be joined to a topological manifold to form a smooth manifold, and a given topological manifold may have several different smooth structures. However, it is not always possible to find any smooth structure for some topological manifolds [1, 7, 8, 11, 19].

Definition 4.3 Let M and N be smooth manifolds and $f : M \to N$ a map. f is *smooth* if, for every $p \in M$, there exist smooth charts (U, φ_U) containing p and (V, φ_V) containing $f(p)$ such that $f(U) \subseteq V$ and the map $\varphi_V \circ f \circ \varphi_U^{-1} : \varphi_U(U) \to \varphi_V(V)$ is smooth.

Definition 4.4 If M and N are smooth manifolds and $f : M \to N$ is a smooth map, for each $p \in M$, the *differential* of f at p, $df_p : T_pM \to T_{f(p)}N$, is defined by

$$df_p(v)(g) = v(g \circ f) \tag{4.1}$$

in which $v \in T_pM$ and $g \in C^\infty(N)$.
df_p is a linear mapping between tangent vector spaces T_pM and $T_{f(p)}N$ [11].

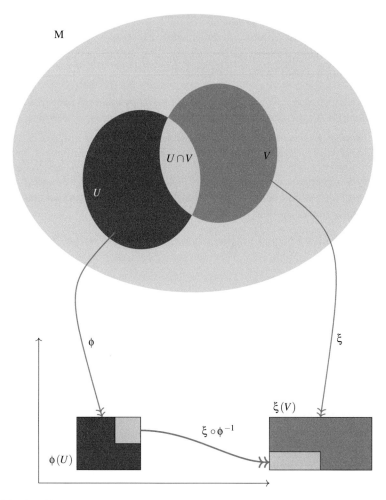

Fig. 4.1 Transition map

Definition 4.5 A smooth map $f : M \rightarrow N$ is a *submersion* when its differential is surjective at each point. It is named an *immersion* if its differential is injective at each point.

Definition 4.6 Let $f : M \rightarrow N$ be a smooth map. $p \in M$ is said to be a regular point of f if $df_p : T_pM \rightarrow T_{f(p)}N$ is surjective. Otherwise, it is named a critical point of f. $y \in N$ is said to be a regular value of f if every element of the level set $f^{-1}(y)$ is regular, and a critical value otherwise.

4.2.2 Some Significant Results

To establish the desired methods, it is necessary to use some well-known results [1, 10, 11] from the field of Global Analysis.

Theorem 1 (Constant rank level set) *Let $f : M \to N$ be a smooth map with constant rank r, where M is an m-dimensional manifold, N is a manifold of dimension n, and $q \in N$ belongs to the image of f. Then the level set $Q \stackrel{\Delta}{=} f^{-1}(q)$ is a properly embedded submanifold of M of dimension $m - r$.*

Corollary 1 (Submersion level set) *If M and N are smooth manifolds and $f : M \to N$ is a smooth submersion, then each level set of f is a properly embedded submanifold whose codimension is equal to the dimension of N.*

Theorem 2 (Sard's Theorem) *Suppose that M and N are smooth manifolds and $f : M \to N$ is a smooth map. Then the set of critical values of f has measure zero in N.*

4.3 The Problem and the Solution

When solving constrained optimization problems, there are usually two types of constraints, that is, equality and inequality restrictions. Certain approaches, like penalty methods, typically incorporate additional terms to the objective function (proportional to the deviations from the feasible regions), trying to maintain evolving candidates inside the desired parts of the overall domain.

In practice, however, this kind of artifice tends to work well for inequality constraints, but equality constraints may produce some problems related to intermittent oscillations around the desired sub-regions during the optimization process evolution. Hence, an almost perfect solution for this phenomenon would be to synthesize candidate points directly inside the feasible subsets and restrict the search to lower dimensional spaces. By taking advantage of this possibility, it would be possible to get at least two benefits—no penalty terms or similar mechanisms due to equality restrictions, and faster search, with fewer dimensions to scan. Although the idea seems reasonable, considering that equality constraints tend to lower the dimension of feasible domains, it is necessary to investigate whether is it possible and under what conditions. Considering Corollary 1, the fact that linear spaces are smooth manifolds, and that most global optimization algorithms are designed to deal with problems defined on Euclidean spaces, it may be interesting to use the Corollary to reach the desired effect. This important line of investigation was already initiated in a certain direction [2, 5, 6] by using concepts and results of Differential Geometry and Topology, in [2] it is focused on optimization of differentiable functions and specific kinds of smooth manifolds.

The objective of this chapter is to present a relatively ambitious idea [3], aimed at eliminating the inconvenient bouncing associated to equality constraints during optimization of not necessarily differentiable functions, defined on smooth manifolds. Such a strategy is to be used together with metaheuristic methods whose candidate populations evolve in linear domains, and intends to reduce the dimension of the underlying search space without losing exploration ability.

Considering that the present method does not interfere with the subset of constraints defined by means of inequalities, it is possible to handle further constraints in an independent way. Furthermore, the framework allows previously approved methods to extend their functionality with tiny changes.

That said, it is time to describe the problem at hand and present the suggested approach to solve it.

4.3.1 The Problem

Let $f : \mathbf{M} \to \mathbb{R}$ be a function, where $\mathbf{M}$ is a finite dimensional smooth n-manifold that is covered by a finite number (N_c) of coordinate domains $\{U_i : i = 1, \ldots, N_c\}$, associated with a finite number of coordinate charts $\{(U_i, \varphi_i) : i = 1, \ldots, N_c\}$. The problem is to find an element $\mathbf{x}^* \in \mathbf{M}$ that globally minimizes f in M. Besides, the solution $\mathbf{x}^*$ has to obey, at least, the following restrictions:

$$h_k(\varphi_*(\mathbf{x}^*)) = 0, k = 1, \ldots, p \qquad (4.2)$$

with $\mathbf{x}^* \in U_*$ and (U_*, φ_*) a coordinate chart in the atlas associated with $\mathbf{M}$. The h_k are smooth real valued functions defined on $\mathbb{R}^n$.

It is assumed that the images of coordinate domains $\varphi_i(U_i)$ are open hyper-rectangles of $\mathbb{R}^n$. This assumption is not a limitation, as most interesting practical situations involve manifolds in which the $\varphi_i(U_i)$ are diffeomorphic to open hyper-rectangles. This condition will make it possible for existing algorihms that originally evolve their populations in the interior of that type of sets to be applied without important changes—of course, other types of (non-evolutionary) global optimization algorithms might be employed as well. Figure 4.2 portrays the desired scheme.

For the sake of simplification, one preliminary measure is necessary: as all images $\varphi_i(U_i) \subset \mathbb{R}^n$ are hyper-rectangles, it is interesting to replace them by only one set, $\mathbf{H} \overset{\triangle}{=} (a_1, b_1) \times (a_2, b_2) \times \cdots \times (a_n, b_n)$, considering that all open hyper-rectangles are pairwise diffeomorphic when $\mathbb{R}^n$ is endowed with its standard topological and differential structures. Accordingly, the corresponding identifying maps (diffeomorphisms between the $\varphi_i(U_i)$ and $\mathbf{H}$) must be composed with the original charts in order to avoid distortions.

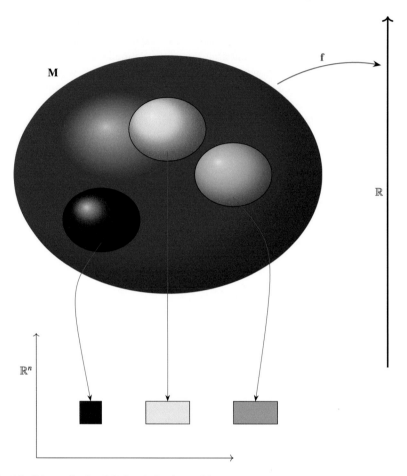

Fig. 4.2 Scheme for the global optimization problem

4.3.2 *The Solution*

Let us discuss here an algorithm for global minimization of real functions defined on a smooth n-manifold $\mathbf{M}$ and satisfying conditions (4.2), assuming the existence of an evolutionary algorithm capable of globally minimizing functions defined on hyper-rectangles of $\mathbb{R}^n$.

Assuming that there exist p equality constraints ($h_k(.) = 0$, $k = 1, \ldots, p$), with $p < n$, we start by defining an auxiliary function $F : \mathbf{M} \to \mathbb{R}^p$ as

$$F(\mathbf{x}) \stackrel{\Delta}{=} (h_1(\varphi_\mathbf{x}(\mathbf{x})), h_2(\varphi_\mathbf{x}(\mathbf{x})), \ldots, h_p(\varphi_\mathbf{x}(\mathbf{x}))) \tag{4.3}$$

in which $\varphi_\mathbf{x}$ is a coordinate map in a proper chart $(U_\mathbf{x}, \varphi_\mathbf{x})$, with $\mathbf{x} \in U_\mathbf{x}$.

Considering the results in Corollary 1, it is easy to see that $F^{-1}(\mathbf{0})$ is a submanifold with codimension p, assuming that $F : \mathbf{M} \to \mathbb{R}^p$ is a smooth submersion, with $\mathbf{0}$ being the origin in $\mathbb{R}^p$. This happens because $\mathbf{M}$ and $\mathbb{R}^p$ are smooth manifolds. On the other hand, $F^{-1}(\mathbf{0})$ is the set of points $\mathbf{x} \in \mathbf{M}$ satisfying $h_k(\varphi_{\mathbf{x}}(\mathbf{x})) = 0, k = 1, \dots, p$. That is, all points satisfying the equality constraints are in $F^{-1}(\mathbf{0})$.

In this setting, dimensional reduction is possible, and it is directly proportional to the number of equality constraints (p), as stated above. Now there are two possible paths to follow:

- Synthesize an atlas for the new submanifold $F^{-1}(\mathbf{0})$ and execute the algorithm indicated below;
- Use a pre-existing optimization method to evolve in subsets of $\mathbb{R}^{n-p}$, fill the gap corresponding to the remaining p components by solving equations defined by the h_k, and compute the objective function values based on the composed point in $\mathbb{R}^n$. These values are to be returned to the calling routine(s).

Therefore, the evolution takes place inside a lower dimensional search space. The algorithm corresponding to the first alternative is shown below.

- **Initialization**
  Presupposing that $F^{-1}(\mathbf{0})$ admits a finite number of coordinate charts $\{(V_i, \Psi_i) : i = 1, \dots, N_s\}$, satisfying all other conditions that hold for the manifold $\mathbf{M}$:

  Construct the expressions for the inverses Ψ_i^{-1} of the coordinate maps Ψ_i ($i = 1, \dots, N_s$), depicted in Fig. 4.3;

  Determine the expressions for the functions $\lambda_i \overset{\Delta}{=} f \circ \Psi_i^{-1} : \Psi_i(V_i) \to \mathbb{R}$ ($i = 1, \dots, N_s$), being $f : \mathbf{M} \to \mathbb{R}$ the original cost function;

  Establish the new cost function as $\beth \overset{\Delta}{=} min\{\lambda_i : i = 1, \dots, N_s\}$.

- **Step 1**
  Complete a new iteration, producing a new population or single point, using $\beth$ values.

- **Step 2**
  If the convergence criteria are met or the maximum number of iterations is reached, go to Final step, else go to Step 1.

- **Final step**
  Calculate $\Psi_i^{-1}(\mathbf{x}^*) \in F^{-1}(\mathbf{0})$, where $\mathbf{x}^*$ is the final product of anterior steps, taken as a global minimizer of $\beth$. Therefore, $\Psi_i^{-1}(\mathbf{x}^*)$ is expected to be a constrained minimizer for f. The index i corresponds to the minimum value $\lambda_i(\mathbf{x}^*)$ for which $\Psi_i^{-1}(\mathbf{x}^*)$ is defined, of course.

It is important to highlight that $\beth$ may be defined in several different ways, depending on the specific problem and the chosen global optimization method. As expected, the convergence properties and performance figures will vary in each possible configuration.

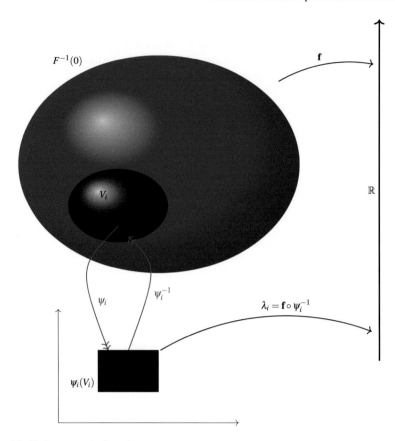

Fig. 4.3 Basic composite functions

In this fashion, all regions of the submanifold $F^{-1}(\mathbf{0})$ will be swept and the resulting populations will lie entirely in it, without using any terms related to equality restrictions. This happens because when evolution takes place, generated points are created directly in that level set. Considering the definition of $\sqsupset$, it is possible to implement the optimization process by means of N_s execution threads, aiming at finding minimizers in each coordinate domain isolatedly and, at the end of each evaluation period, choose the best one—naturally, it is needed to have at our disposal interthread or interprocess communication and syncronization mechanisms. It is obvious that, if the number of charts to which most points belong simultaneously is high, such a procedure tends to be inefficient, but in most practical cases it is certainly a feasible alternative. Moreover, such an approach is a way to eliminate the computational load corresponding to equality constraints in many constrained global optimization problems, and reduce the dimension of the search space.

As an alternative path, and in case the first solution method is not interesting or feasible, it is presented below a second approach, observing that we still have the theoretical guarantee that $F^{-1}(\mathbf{0})$ is a smooth submanifold of codimension p:

- **Preparation**
  - Having assured that $F^{-1}(\mathbf{0})$ is a submanifold of codimension p, adjust the auxiliary method to generate $(n - p)$-dimensional mutually independent points and compute the value of the cost function having only the $(n - p)$-uple as input, that will be complemented as described ahead. The independent parameters must be selected so as to make it possible to obtain the remaining p dependent variables by solving the system formed by the p equations $h_k(\mathbf{x}) = 0, k = 1, \ldots, p$, at each cost function calculation. Of course, the solution for each equation system can be obtained in several ways.
  - Obtain the expressions for the inverses Ψ_i^{-1} of coordinate maps Ψ_i ($i = 1, \ldots, N_c$) of $\mathbf{M}$;
  - Find the analytical expressions for the composite functions $\lambda_i \triangleq f \circ \Psi_i^{-1} :$ $\Psi_i(U_i) \to \mathbb{R}$ ($i = 1, \ldots, N_c$), where $f : \mathbf{M} \to \mathbb{R}$ is the actual cost function. The calculation of the λ_i functions must be preceded by a preprocessing step, during which $(n - p)$ independent input parameters will be used to find p dependent parameters and, finally, the function values will be computed using vectors of $\mathbb{R}^n$;
  - Establish a new cost function by $\aleph \triangleq min\{\lambda_i : i = 1, \ldots, N_c\}$.

- **Step 1**
  Trigger a new iteration of the underlying evolutionary optimization algorithm, generating new candidate(s) driven by $\aleph$ values.

- **Step 2**
  If convergence occurs or the number of iterations has been reached, go to Final step, else go to Step 1.

- **Final step**
  Obtain $\Psi_i^{-1}(\mathbf{x}^*) \in \mathbf{M}$, being $\mathbf{x}^*$ the output of previous steps, expected to be the global minimizer of $\aleph$. Therefore, $\Psi_i^{-1}(\mathbf{x}^*)$ is expected to be a global minimizer of f. As before, the index i corresponds to the minimum value $\lambda_i(\mathbf{x}^*)$ for which $\Psi_i^{-1}(\mathbf{x}^*)$ is defined.

Considering that it is not so easy to construct atlases for arbitrary submanifolds, it seems helpful to present some hints about the synthesis of charts for $F^{-1}(\mathbf{0})$, or the way to properly choose the independent variables to be generated by the auxiliary optimization algorithm. The material presented in [3, 10] is taken as the basis for the following considerations:

Let P be an element of $F^{-1}(\mathbf{0})$. Then it is possible to build a slice chart from a centered chart (V, Ψ) of $\mathbf{0}$ $(= F(P))$ and another one (U, ϕ) of P.

Assuming that $F(U) \subseteq V$ and that the full rank hypothesis is satisfied, it is true that $D(\Psi \circ F \circ \phi^{-1})(\mathbf{0})$ has rank p, and takes the canonical basis $\{\mathbf{e}_1, \mathbf{e}_2, \ldots, \mathbf{e}_n\}$ of $\mathbb{R}^n$ onto a spanning set of $\mathbb{R}^p$—when $n > p$, it is possible to extract a basis from it.

Let us represent such a basis by $\{D(\Psi \circ F \circ \phi^{-1})(\mathbf{0})(\mathbf{e}_{\sigma(1)}), D(\Psi \circ F \circ \phi^{-1})(\mathbf{0})$ $(\mathbf{e}_{\sigma(2)}), \ldots, D(\Psi \circ F \circ \phi^{-1})(\mathbf{0})(\mathbf{e}_{\sigma(p)})\}$, in which $\sigma : \{1, \ldots, p\} \to \{1, \ldots, n\}$ is a strictly increasing function. Moreover, representing the extra $n - p$ unit vectors as $\{\mathbf{e}_{\tau(1)}, \mathbf{e}_{\tau(2)}, \ldots, \mathbf{e}_{\tau(n-p)}\}$, where $\tau : \{1, \ldots, n - p\} \to \{1, \ldots, n\}$ is also a strictly increasing function, we can set $\pi_\tau : \mathbb{R}^n \to \mathbb{R}^{n-p}$ as

$$\pi_\tau(x_1, x_2, \ldots, x_n) \overset{\Delta}{=} (x_{\tau(1)}, x_{\tau(2)}, \ldots, x_{\tau(n-p)}) \tag{4.4}$$

and $\Omega : \phi(U) \to \mathbb{R}^n \cong \mathbb{R}^{n-p} \oplus \mathbb{R}^p$ as

$$\Omega(\mathbf{x}) \overset{\Delta}{=} (\pi_\tau(\mathbf{x}), (\Psi \circ F \circ \phi^{-1})(\mathbf{x})). \tag{4.5}$$

Therefore, the first $n - p$ coordinates correspond to $\pi_\tau(\mathbf{x})$ and the last p to $(\Psi \circ F \circ \phi^{-1})(\mathbf{x})$.

It is possible to prove [10] that Ω is a diffeomorphism whose domain includes $\phi(U_1)$, being $U_1 \subseteq U$ an open neighborhood of P. This fact allows us to conclude that $(U_1, \Omega \circ \phi)$ is a centered chart at it, because $\Omega \circ \phi$ is a diffeomorphism. It is also true that $\Omega(\mathbf{0}) = \mathbf{0}$.

Also, for $\mathbf{x} \in U_1$,

$(\Omega \circ \phi)(\mathbf{x}) = \mathbf{0}$ in the last p coordinates $\iff$

$\Psi(F(\mathbf{x})) = ((\Psi \circ F \circ \phi^{-1})(\phi(\mathbf{x})) = \mathbf{0} \iff F(\mathbf{x}) = \mathbf{0} \iff \mathbf{x} \in F^{-1}(\mathbf{0})$.

So, $(U_1, F \circ \phi)$ is a slice chart at P.

The previous developments demonstrate that it is possible to obtain charts for $F^{-1}(\mathbf{0})$ in an almost programmatic way, making it feasible to obtain atlases if proper starting points are chosen. In addition, we note that the projection $\pi_\tau(.)$ in (4.4) indicates a specific subset of independent variables that will be used by the evolutionary global optimization algorithm to actually make evolution happen.

As aforementioned, the presented method will be illustrated by means of the Fuzzy ASA paradigm, and in what follows some numerical simulations will be presented.

4.4 Examples of Application and Performance Evaluation

In this section many numerical experiments are shown in order to demonstrate the adequacy of the presented ideas. Several constrained global optimization problems defined in [12] and further discussed in [3] are used. The target is to measure the optimization power of the algorithm and illustrate some implementation details. In the first cases the comparisons involve results corresponding to the presented paradigm (Fuzzy ASA with dimensional reduction) and the standard procedure, that employs equality penalties into the original objective function, trying to keep evolving points inside the feasible domain. Please, note that in the former method the generation of new points happens in a region contained in a state space whose dimension is smaller than that in the latter, and the comparison takes into account the number of objective function evaluations necessary to reach convergence. For practical reasons, the tests are based on submanifolds contained in Euclidean spaces and, in computational terms, it is more convenient the use of Jacobian matrices for the sake of investigating the structure of derivatives of vector functions associated to the equality constraints. This is so because these matrices represent the derivatives, as they are linear maps between tangent spaces [11].

An important remark is that in all tests relative to the presented method all equality constraints are satisfied by design, given that generated points are computed already inside the corresponding submanifold.

Therefore, considering that previous results [9, 13, 16, 17, 20] follow different criteria (tolerance of 0.0001 for equality constraints [12]), some minimum values found by means of the present method seem to be suboptimal but, the "official" testing rule [12] accepts a small degree of unfeasibility, leading to an apparent incoherence between results. Actually, it is possible to find several published results that are not completely feasible, in terms of equality constraints.

4.4.1 Problem g03 [3, 12]

This test presents one nonlinear equality constraint and its domain is $[0, 1]^{10} \subset \mathbb{R}^{10}$. The minimum theoretical value for it is -1 at $(10^{-1/2}, 10^{-1/2}, 10^{-1/2}, 10^{-1/2}, 10^{-1/2}, 10^{-1/2}, 10^{-1/2}, 10^{-1/2}, 10^{-1/2}, 10^{-1/2})$. In [12] the minimum value is reported as being -1.00050010001000.

Its precise definition is:

Minimize

$$f(\mathbf{x}) \stackrel{\Delta}{=} -(\sqrt{n})^n \prod_{i=1}^{n} x_i \tag{4.6}$$

subject to

$$h_1(\mathbf{x}) = \sum_{i=1}^{n} x_i^2 - 1 = 0 \quad \text{with } n = 10 \tag{4.7}$$

Here, the conventional procedure would be to search in a 10-dimensional domain contained in $\mathbb{R}^{10}$, but the current algorithm will search on the surface of the

9-dimensional sphere $\mathbb{S}^9$, considering the dimensional reduction due to the arguments decribed above (please, note that in this case there is only one equality constraint).

Let us follow in detail the indications in Sect. 4.3.2 and define $F_1 : \mathbb{R}^{10} \to \mathbb{R}^1$ (refer to Eq. (4.3)) by

$$F_1(\mathbf{x}) \stackrel{\triangle}{=} h_1(\varphi_{\mathbf{x}}(\mathbf{x})) \tag{4.8}$$

in which $\varphi_{\mathbf{x}}$ is the identity map and $U_{\mathbf{x}}$ is $\mathbb{R}^{10}$.

Simplifying, it becomes

$$F_1(\mathbf{x}) = h_1(\mathbf{x}) \tag{4.9}$$

The focus of investigation is $F_1^{-1}(0)$, and the feasible region is contained in $F_1^{-1}(0)$. Hence, by observing that h_1 is smooth, so is F_1, and it is advisable to decide whether it is possible to apply previously stated results about dimensional reduction of the search domain. We thus have to obtain the expression of the Jacobian matrix of F_1 in order to study the variability of its rank, as follows.

$$\begin{aligned}
&J_{F_1}(x_1, x_2, x_3, x_4, x_5, x_6, x_7, x_8, x_9, x_{10}) \\
&= \begin{pmatrix} 2x_1 & 2x_2 & 2x_3 & 2x_4 & 2x_5 & 2x_6 & 2x_7 & 2x_8 & 2x_9 & 2x_{10} \end{pmatrix}
\end{aligned} \tag{4.10}$$

As the premises of Theorem 2 are true, by computing the minors (of order 1) of (4.10) we find that F_1 has rank smaller than 1 inside a set of measure 0 in $\mathbb{R}^{10}$. In this test example there is no feasible point in that region.

Hence, the basic conditions needed to apply the algorithm are true and it is possible to evolve candidate populations directly inside a 9-dimensional region, completing the whole vector just before the computation of the objective function.

After that, the objective function value is returned to the auxiliary algorithm, that restarts the generating process by creating additional 9-dimensional candidate points until the entire process is finished.

In [3], it is reported that after 30 executions of each type of test (proposed and penalized), both converged to the global minimizer in 100 % of the cases, with the difference that the penalized one was slower and less accurate, presenting worse results than the presented here. After repeating similar experiments, the same conclusions were obtained, as can be seen from the comparative graph below, in which average curves corresponding to the several simulations are displayed.

As depicted in Fig. 4.4, it is not difficult to conclude that both techniques are able to find the global minimum, but the present method spends fewer function evaluations to converge. Note that we have different cost functions in each type of execution, occurring that in the penalty based process it is necessary to incorporate the computation of the terms corresponding to the constraints. On the other hand, in the presented method there may be additional function evaluations, relative to the

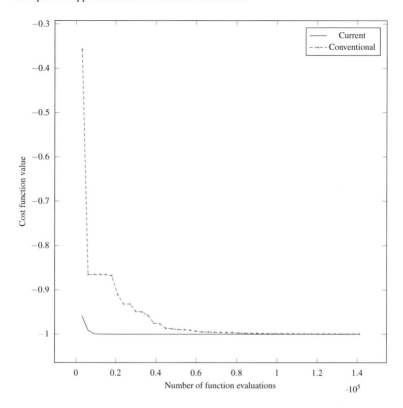

Fig. 4.4 g03 tests

several charts of the manifold. Anyway, the possibility of evolving directly in the natural domain of a given problem is an enormous advantage.

4.4.2 Problem g13 [3, 12]

This example features three nonlinear equality constraints, and the universal domain of the objective function is $[-2.3, 2.3] \times [-2.3, 2.3] \times [-3.2, 3.2] \times [-3.2, 3.2] \times [-3.2, 3.2] \subset \mathbb{R}^5$, and the best known value is 0.053941514041898 at $(-1.71714224003, 1.59572124049468, 1.8272502406271, -0.763659881912867, -0.76365986736498)$ (Ref. [12]).

The formal definition is:

Minimize

$$f(\mathbf{x}) \stackrel{\Delta}{=} e^{x_1 x_2 x_3 x_4 x_5} \tag{4.11}$$

subject to

$$h_1(\mathbf{x}) = \sum_{i=1}^{5} x_i^2 - 10 = 0$$

$$h_2(\mathbf{x}) = x_2 x_3 - 5 x_4 x_5 = 0 \qquad\qquad (4.12)$$

$$h_3(\mathbf{x}) = x_1^3 + x_2^3 + 1 = 0$$

Here, the original problem, if handled in the conventional way, would trigger a search process in a "solid" 5-dimensional domain contained in $\mathbb{R}^5$. However, the current method evolves inside a 2-dimensional submanifold, by virtue of the dimensional reduction (5-dimensional universe with 3 equality constraints).

In order to see the reason why this is so, let us remember the contents of Sect. 4.3.2, defining $F_2 : \mathbb{R}^5 \to \mathbb{R}^3$ by

$$F_2(\mathbf{x}) \stackrel{\Delta}{=} (h_1(\varphi_\mathbf{x}(\mathbf{x})), h_2(\varphi_\mathbf{x}(\mathbf{x})), h_3(\varphi_\mathbf{x}(\mathbf{x}))) \qquad\qquad (4.13)$$

being $\varphi_\mathbf{x}$ chosen as the identity, and $U_\mathbf{x}$, the space $\mathbb{R}^5$.

Substituting, it results in

$$F_2(\mathbf{x}) = (h_1(\mathbf{x}), h_2(\mathbf{x}), h_3(\mathbf{x})) \qquad\qquad (4.14)$$

The main interest is directed to $F_2^{-1}(0, 0, 0)$, expressed by

$$\{(x_1, x_2, x_3, x_4, x_5) \in \mathbb{R}^5 : h_1(x_1, x_2, x_3, x_4, x_5) = h_2(x_1, x_2, x_3, x_4, x_5)$$
$$= h_3(x_1, x_2, x_3, x_4, x_5) = 0\} \qquad\qquad (4.15)$$

Therefore, it is easy to see that the feasible region is contained in $F_2^{-1}(0, 0, 0)$.

As h_1, h_2 and h_3 are C^∞, so is F_2, and it is necessary to investigate more deeply whether is it possible to apply previous results about dimensional reduction of the search domain. The first step is to find the expression for the Jacobian matrix of F_2 in $\mathbb{R}^5$.

$$J_{F_2}(x_1, x_2, x_3, x_4, x_5) = \begin{pmatrix} 2x_1 & 2x_2 & 2x_3 & 2x_4 & 2x_5 \\ 0 & x_3 & x_2 & -5x_5 & -5x_4 \\ 3x_1^2 & 3x_2^2 & 0 & 0 & 0 \end{pmatrix} \qquad\qquad (4.16)$$

Applying Theorem 2 and computing the minors of (4.16), we find that F_2 has rank smaller than 3 inside a set of measure 0 in $\mathbb{R}^5$. Hence, it is possible to apply the presented algorithm in its simpler "flavor", generating candidate points inside a 2-dimensional space and completing the 5-dimensional vector only when computing the objective function.

In this fashion, an iteration is completed by returning the objective function value to the main algorithm, that generates continually new 2-dimensional points until the optimization process is finished.

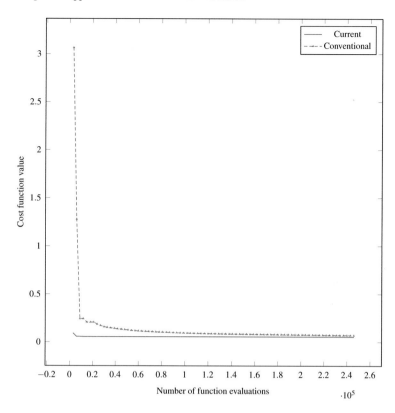

Fig. 4.5 Minimization evolution—g13

As in [3], 30 runs of each type of test were done, and the results confirmed, that is, both methods converged to the global minimizer in all cases, but the penalized was slower, spending much more function evaluations to get to the global minimizer. Figure 4.5 displays the average evolution of runs for each type of method, and allows us to conclude that the presented method is much faster than the penalized one.

4.4.3 Problem g14 [3, 12]

In this problem there are three linear equality constraints and the unconstrained domain is $(0, 10]^{10}$. The minimum known value for this case is -47.7648884594915 ([12]).

The problem definition is:

Minimize

$$f(\mathbf{x}) \overset{\Delta}{=} \sum_{i=1}^{10} x_i (c_i + ln \frac{x_i}{\sum_{j=1}^{10} x_j}) \qquad (4.17)$$

with $c_1 = -6.089, c_2 = -17.164, c_3 = -34.054,$
         $c_4 = -5.914, c_5 = -24.721, c_6 = -14.986,$ (4.18)
         $c_7 = -24.1, c_8 = -10.708, c_9 = -26.662, c_{10} = -22.179$

subject to

$$h_1(\mathbf{x}) = x_1 + 2x_2 + 2x_3 + x_6 + x_{10} - 2 = 0$$
$$h_2(\mathbf{x}) = x_4 + 2x_5 + x_6 + x_7 - 1 = 0$$
$$h_3(\mathbf{x}) = x_3 + x_7 + x_8 + 2x_9 + x_{10} - 1 = 0$$ (4.19)

where $\mathbf{x} = (x_1, x_2, x_3, x_4, x_5, x_6, x_7, x_8, x_9, x_{10}) \in \mathbb{R}^{10}$

As before, the conventional solution for this problem would be conducted by a search process over a 10-dimensional domain, whereas the presented algorithm evolves in a 7-dimensional submanifold, thanks to the existence of three equality constraints. In order to make this statement precise, let $F_3 : \mathbb{R}^{10} \to \mathbb{R}^3$ be defined by

$$F_3(\mathbf{x}) \triangleq (h_1(\varphi_{\mathbf{x}}(\mathbf{x})), h_2(\varphi_{\mathbf{x}}(\mathbf{x})), h_3(\varphi_{\mathbf{x}}(\mathbf{x})))$$ (4.20)

being $\varphi_{\mathbf{x}}$ the identity map, and $U_{\mathbf{x}}$ is $\mathbb{R}^{10}$.

Hence, the expression simplifies to

$$F_3(\mathbf{x}) = (h_1(\mathbf{x}), h_2(\mathbf{x}), h_3(\mathbf{x}))$$ (4.21)

In this example, the aim is to investigate the properties of $F_3^{-1}(0, 0, 0)$, that is nothing more than $\{\mathbf{x} \in \mathbb{R}^{10} : h_1(\mathbf{x}) = h_2(\mathbf{x}) = h_3(\mathbf{x}) = 0\}$. As h_1, h_2 and h_3 are smooth, so is F_3, and to use the previous results it is necessary to establish some additional facts. To that end, the rank of the Jacobian matrix of F_3 at a generic point of $\mathbb{R}^{10}$ must be studied. Its expression is

$$J_{F_3}(x_1, x_2, x_3, x_4, x_5, x_6, x_7, x_8, x_9, x_{10}) = \begin{pmatrix} 1 & 2 & 2 & 0 & 0 & 1 & 0 & 0 & 0 & 1 \\ 0 & 0 & 0 & 1 & 2 & 1 & 1 & 0 & 0 & 0 \\ 0 & 0 & 1 & 0 & 0 & 0 & 1 & 1 & 2 & 1 \end{pmatrix}$$ (4.22)

As the conditions for application of Theorem 2 are true and (4.22) has full rank in the domain of F_3, we can apply the algorithm in its simpler form by generating points inside a 7-dimensional space and composing the full vector immediately before the actual computation of the cost function. As usual, each iteration is completed when the cost function value is returned to the driving module. This cyclic process repeats until the finishing criteria are met.

In this case study (as in [3]), after 30 optimization sessions for each type of simulation (presented and penalized), both methods approached a small neighborhood of the assumed global minimizer, but only the proposed method truly converged within the chosen maximum number of function evaluations. In Fig. 4.6 it is shown a graph portraying the evolution of the tests.

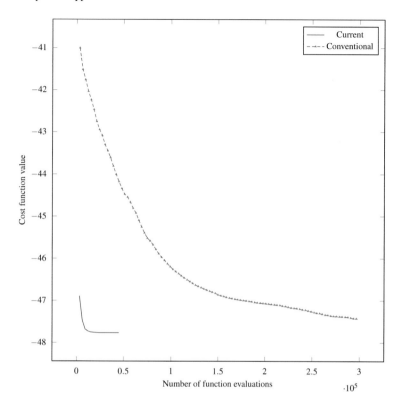

Fig. 4.6 Minimization evolution—g14

4.4.4 Problem g15 [3, 12]

This test features two equality constraints and the domain is $[0, 10]^3$.

The minimum value is 961.715022289961 ([12]).

The formal definition is given by:

Minimize

$$f(\mathbf{x}) \overset{\Delta}{=} 1000 - x_1^2 - 2x_2^2 - x_3^2 - x_1 x_2 - x_1 x_3 \qquad (4.23)$$

subject to

$$
\begin{aligned}
h_1(\mathbf{x}) &= x_1^2 + x_2^2 + x_3^2 - 25 = 0 \\
h_2(\mathbf{x}) &= 8x_1 + 14x_2 + 7x_3 - 56 = 0
\end{aligned}
\qquad (4.24)
$$

As usual, the conventional search process takes place inside $\mathbb{R}^3$. However, the new algorithm evolves in a 1-dimensional interval, considering the dimensional reduction made possible by the existence of 2 equality constraints.

Let us start by defining $F_4 : \mathbb{R}^3 \to \mathbb{R}^2$ (relative to Eq. (4.3)) as

$$F_4(\mathbf{x}) \overset{\Delta}{=} (h_1(\varphi_{\mathbf{x}}(\mathbf{x})), h_2(\varphi_{\mathbf{x}}(\mathbf{x}))) \tag{4.25}$$

in which $\varphi_{\mathbf{x}}$, again chosen as the identity map, and $U_{\mathbf{x}}$ as $\mathbb{R}^3$.

So, we get the expression

$$F_4(\mathbf{x}) = (h_1(\mathbf{x}), h_2(\mathbf{x})) \tag{4.26}$$

Now, the target is $F_4^{-1}((0,0))$, defined by $\{(x_1, x_2, x_3) \in \mathbb{R}^3 : h_1(x_1, x_2, x_3) = h_2(x_1, x_2, x_3) = 0\}$.

Being h_1 and h_2 smooth, so is F_4, making it possible the application of the results about dimensional reduction of the optimization domain. Exactly as in other examples, it is necessary to compute the expression for the Jacobian matrix of F_4 at a generic point of $\mathbb{R}^3$. It is given by

$$J_{F_4}(x_1, x_2, x_3) = \begin{pmatrix} 2x_1 & 2x_2 & 2x_3 \\ 8 & 14 & 7 \end{pmatrix} \tag{4.27}$$

Observing that Theorem 2 may be applied, and analizing the minors of (4.27), we find that F_4 has rank smaller than 2 only in a set of measure 0 in $\mathbb{R}^3$. Furthermore, $F_4^{-1}((0,0))$ meets that set in one point. In practice, the probability of generating such a point is zero, and such event may be considered impossible. Anyway, even when the impossible happens, it is possible to discard the point and continue the processing. Repeating the anterior procedure, that is, the simpler "flavor" of the proposed algorithm, the evolution will happen inside a 1-dimensional region, with the construction of the 3-dimensional vector occurring just before the actual computation of the objective function. Following this reasoning, we arrive to Eq. (4.28),

$$245x_2^2 + (224x_1 - 1568)x_2 + 113x_1^2 - 896x_1 + 1911 = 0 \tag{4.28}$$

making it possible to find x_2 as a function of x_1, here chosen as the independent variable. After obtaining the two first coordinates, x_3 is found by means of other constraint defining relations.

Eventually the objective function value is returned to the controlling algorithm, which generates further 1-dimensional candidate points until the whole procedure is finished.

Finally, it is necessary to assess the quality of the proposed algorithm when applied to this problem. Following the test procedure established in [3], and after 30 simulations for each type of test, the presented method converged to the global minimizer in 100% of the cases, while the penalized one was not able to offer the same level.

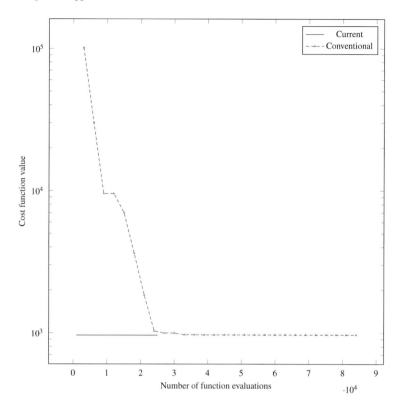

Fig. 4.7 Minimization evolution—g15

In Fig. 4.7 we can find a graph showing the average evolution of all runs, relatively to each type of method.

4.4.5 A Different Viewpoint

Following Ref. [3], further problems from the CEC 2006 Special Session on Constrained Real-Parameter Optimization test suite are used to better evaluate the presented method. So, problems g11, g21 and g23, defined below, were chosen. These functions complement the previous subset in that they feature at least one equality constraint, that is our essential condition to open the possibility to reduce the dimension of search domains.

It is worth to warn the reader that in the following simulations the rule prescribed in [12] holds, that is, all information related to function evaluations refer to those needed in each run for finding an approximation satisfying the condition

$|f(\mathbf{x}) - f(\mathbf{x}^{optimal})| < 0.0001$, differently from the previous examples. So, the testing approaches are different, because now the simulations are interrupted earlier.

In the following tests, and obeying the rules established in [12], each simulation is repeated 25 times under exactly the same conditions. In each case, the evolution of minimization sessions are shown in order to make it easier to interpret the obtained results. Differently from [3], the comparisons involve the conventional and the current methods coupled to Fuzzy ASA, and results are reported by means of graphs of all minimization runs.

4.4.5.1 Definition and Results for Problem g11 [3, 12]

Problem g11 is defined by:

Minimize

$$f(\mathbf{x}) = x_1^2 + (x_2 - 1)^2 \tag{4.29}$$

subject to

$$h_1(\mathbf{x}) = x_2 - x_1^2 = 0 \tag{4.30}$$

The unconstrained domain of the cost function is $[-1, 1]^2 \subset \mathbb{R}^2$, and the problem presents only one nonlinear equality constraint. The minimum value for this problem is 0.75 at $(\pm\sqrt{2}/2, 0.5)$.

Figure 4.8 displays comparative graphs that allow us to conclude that the presented method is really effective.

4.4.5.2 Definition and Results for Problem g21 [3, 12]

Problem g21 may be stated as:

Minimize

$$f(\mathbf{x}) = x_1 \tag{4.31}$$

subject to

$$
\begin{aligned}
g_1(\mathbf{x}) &= -x_1 + 35x_2^{0.6} + 35x_3^{0.6} \le 0 \\
h_1(\mathbf{x}) &= -300x_3 + 7500x_5 - 7500x_6 - 25x_4x_5 + 25x_4x_6 + x_3x_4 = 0 \\
h_2(\mathbf{x}) &= 100x_2 + 155.365x_4 + 2500x_7 - x_2x_4 - 25x_4x_7 - 15536.5 = 0 \\
h_3(\mathbf{x}) &= -x_5 + ln(-x_4 + 900) = 0 \\
h_4(\mathbf{x}) &= -x_6 + ln(x_4 + 300) = 0 \\
h_5(\mathbf{x}) &= -x_7 + ln(-2x_4 + 700) = 0
\end{aligned}
\tag{4.32}
$$

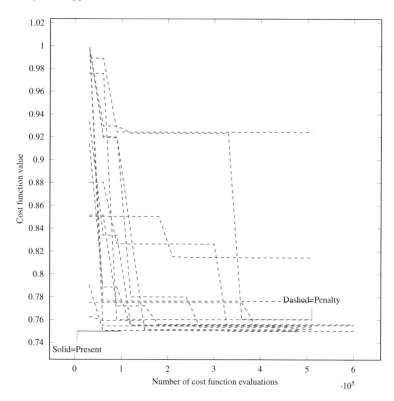

Fig. 4.8 g11—comparison between 2 methods

The unconstrained domain of this function is $[0, 1000] \times [0, 40] \times [0, 40] \times [100, 300] \times [6.3, 6.7] \times [5.9, 6.4] \times [4.5, 6.25] \subset \mathbb{R}^7$, whereas the feasible domain has to obey further five nonlinear equality constraints. The minimum known value for this problem ([12]) is 193.724510070034967 at (193.724510070034967, 5.56944131553368433e-27, 17.3191887294084914, 100.047897801386839, 6.68445185362377892, 5.99168428444264833, 6.21451648886070451).

Figure 4.9 displays comparative graphs—again, the presented method converged very fast, when compared to the conventional approach.

4.4.5.3 Definition and Results for Problem g23 [3, 12]

This example is defined by:

Minimize

$$f(\mathbf{x}) = -9x_5 - 15x_8 + 6x_1 + 16x_2 + 10(x_6 + x_7) \qquad (4.33)$$

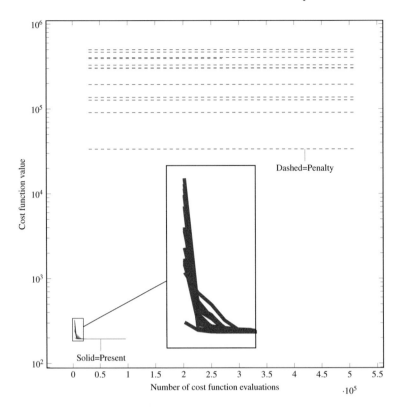

Fig. 4.9 g21—comparison between 2 methods

subject to

$$
\begin{aligned}
g_1(\mathbf{x}) &= x_9 x_3 + 0.02 x_6 - 0.025 x_5 \le 0 \\
g_2(\mathbf{x}) &= x_9 x_4 + 0.02 x_7 - 0.015 x_8 \le 0 \\
h_1(\mathbf{x}) &= x_1 + x_2 - x_3 - x_4 = 0 \\
h_2(\mathbf{x}) &= 0.03 x_1 + 0.01 x_2 - x_9(x_3 + x_4) = 0 \\
h_3(\mathbf{x}) &= x_3 + x_6 - x_5 = 0 \\
h_4(\mathbf{x}) &= x_4 + x_7 - x_8 = 0
\end{aligned}
\tag{4.34}
$$

The irrestricted domain of the cost function is $[0, 300] \times [0, 300] \times [0, 100] \times [0, 200] \times [0, 100] \times [0, 300] \times [0, 100] \times [0, 200] \times [0.01, 0.03] \subset \mathbb{R}^9$, being the minimum known value for it assumed [12] as -400.055099999999584 at $(0.00510000000000259465, 99.9947000000000514, 9.01920162996045897e-18, 99.9999000000000535, 0.000100000000027086086, 2.75700683389584542e-14, 99.9999999999999574, 200, 0.0100000100000100008)$.

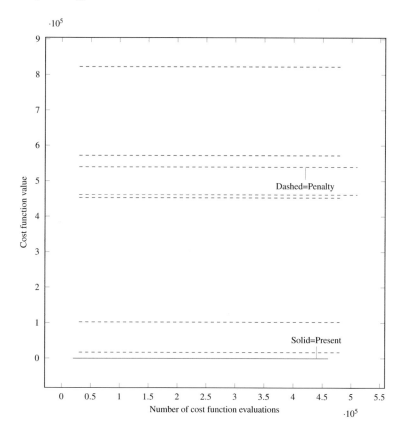

Fig. 4.10 g23—comparison between 2 methods

On the other hand, it is possible to reach the point (0.0, 100.0, 0.0, 100.0, 0.0, 0.0, 100.0, 200.0, 0.01) without violating any constraint, with the cost function value equal to −400.00—at this point all inequality constraints are active.

Therefore, we assume that this is the exact solution for problem g23.

Figure 4.10 displays comparative graphs allowing us to conclude, once more, that the current approach is effective in comparison with the penalty method.

4.4.5.4 Interpretation

As can be seen from the graphs, the presented method shows better performance when compared to the conventional paradigm. Such a striking difference may be attributed to the dimensional reduction of the search domain and, although the original problem is the same, the new method evolves in less dimensions than the penalized. The new configuration space is lower dimensional than the initial one and, consequently, the search tends to be faster than it would be in the containing manifold. Needless to say that this is possible when the constraining expressions satisfy all necessary prerequisites.

4.5 Conclusions

This chapter introduced an alternative approach to constrained global optimization on smooth manifolds that allows, among other things, to handle functions whose natural domains are not linear—the evolution of candidate points happens directly in regions defined by equality constraints. Thanks to its very flexible structure, the method allows coupling to many already existing evolutionary methods that can take advantage of its structure. The numerical results lead us to conclude that the presented method is effective and able to eliminate problems associated to techniques aiming at the same target, like imposing penalty terms relative to equality constraints so as to force evolving populations to keep inside specific regions of the search space. Furthermore, previous algorithms are comparatively complex, considering that constructing optimization algorithms in the Riemannian manifold environment is certainly not an easy task. On the other hand, the alternative proposal is able to deal with nonsmoothness in a simple way, thanks to the inherent characteristics of metaheuristics. Also, the presented procedures make it possible to explain precisely the common practice of generating only part of the coordinates of evolving points and complete the full candidate vector by solving the equations that describe the constraints just before the evaluation of the objective function—this property is very important. Finally, it is important to reinforce that the presented ideas can also be useful when coupled to other algorithms like several DE, PSO, ABC or GA implementations [14, 18, 21].

References

1. Abraham, R., Marsden, J.E., Ratiu, T.: Manifolds, Tensor Analysis, and Applications. Springer, New York (1988)
2. Absil, P.-A., Mahony, R., Sepulchre, R.: Optimization Algorithms on Matrix Manifolds. Princeton University Press, Princeton (2008)
3. Aguiar e Oliveira Jr, H., Petraglia, A.: Dimensional reduction in constrained global optimization on smooth manifolds. Inf. Sci. **299**, 243–261 (2015)
4. Chern, S.S., Shen, Z.: Riemann-Finsler Geometry. World Scientific, Singapore (2005)
5. Helmke, U., Moore, J.B.: Opti. Dyn. Syst. Springer, London (1994)
6. Hillermeier, C.: Nonlinear Multiobjective Optimization - A Generalized Homotopy Approach. Birkhäuser Verlag, Basel (2001)
7. Hirsch, M.W.: Differential Topology. Springer, New York (1976)
8. Kock, A.: Synthetic Geometry of Manifolds. Cambridge University Press, New York (2009)
9. Kukkonen, S., Lampinen, J.: Constrained real-parameter optimization with generalized differential evolution. In: CEC 2006, Vancouver, pp. 207–214 (2006)

10. Lawson, J.: Calculating Submanifold Charts, Lecture Notes, Department of Mathematics, Louisiana State University, Louisiana. https://www.math.lsu.edu/~lawson/subman06.pdf. Accessed 24 March 2015
11. Lee, J.M.: Introduction to Smooth Manifolds, 2nd edn. Springer, New York (2013)
12. Liang, J.J., Runarsson, T.P., Mezura-Montes, E., Clerc, M., Suganthan, P.N., Coello, C.A.C., Deb, K.: Problem definitions and evaluation criteria for the CEC 2006 special session on constrained real-parameter optimization. Technical Report, Nanyang Technological University, Singapore (2005)
13. Liang, J. J., Suganthan, P. N.: Dynamic Multi–Swarm Particle Swarm Optimizer with a Novel Constraint–Handling Mechanism. In: CEC 2006, Vancouver, pp. 9–16 (2006)
14. Long, Q.: A constraint handling technique for constrained multi-objective genetic algorithm. Swarm Evol. Comput. **15**, 66–79 (2014)
15. Ma, Y., Fu, Y.: Manifold Learning Theory and Applications. CRC Press, Boca Raton (2012)
16. Mezura-Montes, E., Velázquez-Reyes, J., Coello Coello, C. A.: Modified differential evolution for constrained optimization. In: CEC 2006, Vancouver, pp. 25–32 (2006)
17. Tasgetiren, M. F., Suganthan, P. N.: A multi-populated differential evolution algorithm for solving constrained optimization problem. In: CEC 2006, Vancouver, pp. 33–40 (2006)
18. Tsai, H.C.: Integrating the artificial bee colony and bees algorithm to face constrained optimization problems. Inf. Sci. **258**, 80–93 (2014)
19. Tu, L.W.: An Introduction to Manifolds. Springer, New York (2011)
20. Zhang, M., Luo, W., Wang, X.: Differential evolution with dynamic stochastic selection for constrained optimization. Inf. Sci. **178**, 3043–3074 (2008)
21. Zhang, G., Cheng, J., Gheorghe, M., Meng, Q.: A hybrid approach based on differential evolution and tissue membrane systems for solving constrained manufacturing parameter optimization problems. Appl. Soft Comput. **13**(3), 1528–1542 (2013)

Part III
Further Applications of Fuzzy ASA

Chapter 5
Nash Equilibria of Finite Strategic Games and Fuzzy ASA

Abstract In this chapter various significant results obtained by means of the application of the Fuzzy Adaptive Simulated Annealing (Fuzzy ASA) algorithm are introduced—the aim is to find all Nash equilibria of finite normal form games. To get there, Fuzzy ASA has been modified in order to incorporate techniques, based on space-filling curves, able to find adequate starting points—several well-known strategic games are used to test the efficacy of the method. The obtained results are compared to previously published results that used similar techniques in order to solve the same problem but could not find all equilibria in all tests. As it is very important to study and model the interactions between agents, the Nash equilibrium concept is widely recognized as a powerful tool, adequate to discover situations in which joint strategies are optimal in the sense that players cannot benefit from changing unilaterally their strategies. In this fashion, any technique that may represent a true advancement, in terms of efficacy when finding whole sets of solutions for a given strategic game, is worth to invest in.

5.1 Introduction

Basically game theory techniques are aimed at modeling the evolution of the interaction between (socio-economic) agents or players, whose actions may influence the course of the whole dynamic process—in many instances the action taken by a player can change the flow of the overall game. The fundamental premises usually are that agents will pursue certain targets (modeled by payoff functions), using their a priori expectations about the other players' possible decisions in order to get to their objectives. Game theory has been used in many sectors, like computer science, management, and economics, even today its most frequent source of problems.

As it is an important instrument aimed at analyzing a huge range of interactions in which the final results depend on the joint strategies taken by many players, the equilibrium concept is considered a good model for a stable outcome of a given game. Although there exist many proposals for the equilibrium notion in the literature, the Nash equilibrium is considered the most important of all, and probably the most studied one.

© Springer International Publishing Switzerland 2016
H. Aguiar e Oliveira Junior, *Evolutionary Global Optimization,*
Manifolds and Applications, Studies in Systems, Decision and Control 43,
DOI 10.1007/978-3-319-26467-7_5

The concept of Nash equilibrium may be defined as a particular configuration of strategies in which each player's choice is an optimal reaction to the other players' strategies [12, 13] in the sense that no one may improve its position in an isolated way. So, the concept establishes a distinguished type of steady state of strategic games in which each player acts rationally and obtains correct expectations with respect to the behavior of other players.

Considering that the determination of Nash equilibria of large normal form games is a nontrivial problem, several methods have been discovered aiming to calculate them [12, 13, 15], with many recent works solving related problems [5–7, 9–11]. One important tool designed to solve finite games is the Gambit package [14], that contains a programming library and associated software tools for the construction and analysis of extensive and normal form games—some of the example games distributed with Gambit were used as test cases to evaluate the method described in [3], that we describe here. The optimal choice of a method for computing Nash equilibria of finite games depends on several factors, as whether you need to find pure or mixed strategy equilibria, or want to find only one or all equilibria, for example. Hence, a solution can be understood as a systematic description of the outcomes that may result from a game [15], and this chapter considers only strategic, normal form games.

As described in [3] and in a previous chapter, the optimization method used in this chapter (Fuzzy ASA) may be thought of as an evolution of the original ASA method [8], having in mind that despite its ample usefulness, the latter did not approach the global optima of certain cost functions used in the process of solving some examples. Besides the improvements present in Fuzzy ASA, in [2] some further improvements were incorporated, in terms of obtaining better starting points in a deterministic fashion. The modification is based on the use of α-dense curves, that are approximations of abstract space-filling curves. In [1] another different, but related, method was introduced, aimed again at finding good departure points in a deterministic way, now using space-filling curves. In that paper it was shown how to synthesize new space-filling curves based upon key theorems of Topology, and how to implement the corresponding approximations. The objective here is to apply this more complex implementation in order to solve the harder problem, that is, finding all Nash equilibria of generic finite strategic games in normal form. In the literature the problem of finding Nash equilibria of finite strategic games was shown to admit different alternative formulations. Nevertheless, computing such solutions remains a difficult task and, furthermore, algorithms unable to compute several (or all) Nash equilibria can be unsatisfactory for many practical tasks.

Often, the central problem of this chapter (finding Nash equilibria of finite normal form games) can be solved by finding fixed points of certain functions [13], but one of the most important theoretical results presently known is that the problem of computing a Nash equilibrium can be faced as a global optimization task [12, 13, 16]. This approach opens up a new horizon of feasible methods of solution for the problem at hand, considering the large number of optimization methods available nowadays, especially those using evolutionary techniques, such as genetic algorithms, particle swarm optimization, or differential evolution.

This chapter is aimed at demonstrating that Fuzzy ASA is effective in the approximation of Nash equilibria for finite strategic games, and capable to overcome other effective methods [16] when searching for all solutions to a given game. To succeed in this task, some test games originally used in [16], and after in [3], are employed here as well, so as to make it possible to compare the different methods. The principal target is to show that the presented method can find all equilibria in all simulations. Here, a multistart version of Fuzzy ASA is used so as to provide a set of starting points with potential to reach all global minima and escape from suboptimal regions—this deterministic preprocessing stage is based on space-filling curves. In what follows the many components of the method will be described, including the application of Fuzzy ASA to the underlying problem and the obtained results.

5.2 Space-Filling Curves and Their Effect on Fuzzy ASA

As happens to other numerical optimization methods that depend on good starting points to improve their performance, ASA and Fuzzy ASA can improve their results when departing from adequate regions. Therefore, it seems interesting to add a preprocessing step that could produce a small set of good seeds, trying to prevent convergence to suboptimal points and making it possible to obtain multiple global minimizers in single simulations. So, a preprocessing scheme, based on space-filling curves, was recently proposed and is explained in references [1, 4]. Since its efficacy depends on its ability to reach all regions of state spaces, it is worth mentioning that, although there are several space-filling curves able to filling up multidimensional domains, at least in theory, the corresponding numerical approximations can exhibit "holes" because of truncation or limitations of present day computing paradigms when dealing with real numbers. It is important to highlight that space-filling curves are able to "visit" deterministically all points of high-dimensional domains, increasing the chance of finding good starting points for other optimization phases. Of particular importance are those located in attraction basins of global optima.

As said before, although this front end may be coupled to any optimization method, here we use the Fuzzy ASA algorithm, considering its good performance in difficult optimization tasks and its robustness.

Here a space-filling curve is understood as a surjective and continuous function from a real interval to a compact subset of a finite-dimensional vector space, which can be identified with $\mathbb{R}^n$. Space-filling curves have been studied through the decades and there exist well-established results stating necessary conditions for their existence [17]. Furthermore, decades before the creation of digital computers, mathematicians constructed examples and established several profound properties of space-filling curves. Nowadays, many additional ways to apply previous knowledge about these curves are known and have been applied to several areas, including optimization of numerical functions.

In this chapter the proposal is to compose a given objective function with a space-filling curve corresponding to a compact set containing the associated domain. By doing so, it is possible to convert a multivariate problem into a univariate one. Therefore, it is possible, by solving the auxiliary one–dimensional problem, to return to the n–dimensional domain and find the desired optimum point. One problem we have to cope with is that such ideas are hard to implement because some difficulties arise, mainly when dealing with high-dimensional spaces. One big obstacle regarding implementation is that many curves proposed in the past are difficult to approximate in digital computers with finite word length, considering that this limitation may cause divergence after some iterations.

5.3 Game Theory and Nash Equilibria—Basic Definitions

5.3.1 Finite Strategic Games

A finite strategic game may be defined as a model for the interaction between agents, or decision makers, which are usually referred to as players. During each move, a player decides which alternative to take, among a pre-established set of possible actions. Such a model describes interactions between the agents by means of functions that measure the mutual effect of the actions of all players and their returns (payoff functions). Formally, a strategic game can be defined in the following way:

A finite strategic game with pure strategies is composed of three elements:

- A finite set $\{1, 2, \ldots, N\}$ of players.
- A set of sets of actions S_i (pure strategies) so that each player is associated to a finite group of strategies which can be chosen. So, we have $S_i = \{s_{i_1}, \ldots, s_{i_{m_i}}\}$, where m_i is the number of actions available to player i.
  In addition, a_{-i} denotes the profile $(a_1, \ldots, a_{i-1}, a_{i+1}, \ldots, a_N)$, obtained from $(a_1, \ldots, a_{i-1}, a_i, a_{i+1}, \ldots, a_N)$ by excluding the action of player i. The notation (a_i, a_{-i}) is used to represent the complete profile of actions.
- A set of payoff functions, one for each player, as follows

$$u_i : S \to \mathbb{R}$$

where
$$S = S_1 \times S_2 \times \cdots \times S_N.$$

These functions take configurations of actions to real values.

Strategic games can model many practical situations, and in several cases of interest, players represent firms, the actions are possible or feasible acquisitions, and the payoff functions the resulting profits relative to the decisions made. Another setting may be a set of competing football teams trying to hire one player, among several other possible alternatives, and the payoff is the final score in the championship. For the type of game studied here, time has no significance, so that the players make decisions once and for all, without relative time delays or similar things, meaning that no player knows about the actions taken by other players—they are simultaneous move games. On the other hand, in certain games an action may include activities that take some time, and might have to take into account a nontrivial set of conditions. The fact that time has no influence in the model means that when analyzing a given phenomenon as a strategic game, we do not take into account the complications that arise if players are allowed to change their viewpoints as events unfold: it is assumed that actions are chosen in a definitive way.

5.3.2 Nash Equilibrium—Fundamental Definitions

A fundamental issue in game theory is the prediction of what decisions will be taken by players in a specific configuration. In general, it is assumed that each agent chooses the best available strategy. The best action for any given player typically depends on the other players' choices, and when choosing a move, a player must consider the positions that the other players can take. In other words, they must establish a belief about the other players' decisions. Therefore, it is worth to figure out in what circumstances can such a belief be built. So, it is assumed that each player can estimate a certain degree of belief from the past experience with the game, and that this previous knowledge is sufficient to provide approximate information about how the agents will behave—this ability will make it possible to enrich the model with mixed strategies. Although nobody tells a particular player the decisions the opponents will make, the past experiences with the game lead players to be sure of these actions. Hence, the common mechanism for solving this problem is based on each player choosing a specific action according to a rational behavior, considering the belief about the other players' behavior, supposing that every player's belief about the other players' decisions is right. The conventional definition of a Nash equilibrium contains these two components, that can be informally stated by saying that it is an action profile with the property that no player can attain better positions by choosing unilaterally an action distinct from the present configuration. Hence, when the action profile gets to a Nash equilibrium, no player has a rational incentive for choosing any action different from the current position—there is no advantage in changing. From another viewpoint, Nash equilibria represent stable social settings, meaning that if all the players converge to it, no one wishes to deviate from that relatively comfortable situation. Another fundamental tenet of the theory is that players' beliefs about each other's choices are correct. So it happens that two players' beliefs about a third player's action tend to be identical—as a result, it is often said that the players' expectations are coordinated. In many cases the environment to which we would like to apply the theory of Nash equilibrium do not correspond exactly to an idealized setting. So, in some cases players might not have enough experience with the game, and in others they may not view each instance of the game in an adequate way. Therefore, the concept of Nash equilibrium may be appropriate in many opportunities but, of course, there are times in which it will be better to choose another paradigm to represent a specific type of equilibrium.

At this point we establish the terms and symbols that will be used along the chapter [16]. $\clubsuit_i$ is the symbol for the set of probability mass functions on S_i, and $\clubsuit$ represents the Cartesian product $\prod_{i=1}^{N} \clubsuit_i$. Hence, we have $\clubsuit \subset \mathbb{R}^m (m = \sum_{i=1}^{N} m_i)$ and the elements of $\clubsuit_i$ belong to finite domains, being real valued functions defined on S_i ($p_i : S_i \to \mathbb{R}$). Furthermore, they need to satisfy

$$\sum_{s_{ij} \in S_i} p_i(s_{ij}) = 1$$

where $p_i(s_{ij}) \geq 0$.

The term s_{ij} represents strategies $p_i \in \clubsuit_i$ satisfying $p_i(s_{ij}) = 1$. In other words, it is a representation for pure strategy profiles. So, (s_{ij}, p_{-i}) is the state in which player i chooses pure strategy s_{ij} (100 % of belief).

The introduction of mixed strategies is necessary to allow the degrees of belief to join the model. In this manner, the p_i allow the players to assign different weights to each possible action of all players by adjusting them, making it possible to calculate an extended and aggregate payoff function (for each player i) that is a kind of expected value for the configurations of the game. Such a function is defined below for player i:

$$\mathbf{U_i}(p) \overset{\Delta}{=} \sum_{s \in S} p(s) u_i(s) \tag{5.1}$$

where

$$p(s) = \prod_{i=1}^{N} p_i(s_i) \tag{5.2}$$

and

$$s = (s_1, s_2, \ldots, s_n) \tag{5.3}$$

At this point it is possible to formulate a definition for a Nash equilibrium. A mixed strategy profile $p^* = (p_1^*, p_2^*, \ldots, p_N^*) \in \clubsuit$ is a *Nash equilibrium* if

$$\mathbf{U_i}(p_i, p_{-i}^*) \leq \mathbf{U_i}(p^*)$$

for all $i = 1, \ldots, N$ and $p_i \in \clubsuit_i$.

Accordingly, for a strategy profile p^* to be a Nash equilibrium it is needed that no player i have an available action that may result in a better payoff than the received by choosing p_i^*, assuming that the other players keep their positions fixed as p_j^* ($j \neq i$). That is, in that special configuration any isolated deviation is prejudicial.

5.3.3 Establishing the Objective Function

It is known [12] that Nash equilibria of normal form games can be found using global minimization of certain real valued functions. According to this line of reasoning, three auxiliary matrix functions ($\mathbf{x}$, $\mathbf{z}$ and $\mathbf{g}$) have to be determined in order to give rise to an overall objective function, whose global minima furnish the Nash equilibria for the given game. Such functions are the following:

$$x_{ij}(p) \stackrel{\Delta}{=} \mathbf{U_i}(s_{ij}, p_{-i}) \tag{5.4}$$

$$z_{ij}(p) \stackrel{\Delta}{=} x_{ij}(p) - \mathbf{U_i}(p) \tag{5.5}$$

$$g_{ij}(p) \stackrel{\Delta}{=} max(z_{ij}(p), 0) \tag{5.6}$$

with $p \in \clubsuit$.

The objective function we are searching for is given by

$$v(p) \stackrel{\Delta}{=} \sum_{i=1}^{N} \sum_{j=1}^{m_i} (g_{ij}(p))^2, \ p \in \clubsuit. \tag{5.7}$$

It is worth to note that p^* is a Nash equilibrium if and only if it is a global minimum of v in $\clubsuit$ (refs. [12, 13, 16]), or $v(p^*) = 0$, considering that $v(p) \geq 0$.

The aim of this chapter is to present a method to cope with the problem of finding all global minima of the function v, defined by (5.7). We follow mainly the ideas presented in [3].

5.4 Solution Based on Fuzzy ASA and Evaluation of Results

In order to assess the usefulness of the presented method the results will be compared to those presented in [16], but the evaluation will be concentrated on the efficacy of the methods, that is, a given method will be labeled as fully effective if and only if it is able to find all minimizers of (5.7) in 100 % of the simulations, even the ones based on stochastic approaches. Proceeding along this line, the same test set is used here in order to compare the two algorithms in a coherent way. Six games compose the test set, and the performance of the algorithms is assessed in all of them. They are included in the GAMBIT software package [14], which can be freely downloaded from http://www.gambit-project.org.

In all tests there are more than one Nash equilibria, and in all cases the fundamental goal is to detect all equilibrium points. In order to obtain all Nash equilibria of each game, it is possible to the use the GAMBIT package. Below, we offer the definition for the test problems along with their equilibria and payoff functions, for the sake of helping the reader when evaluating the results. Furthermore, to each game corresponds a GAMBIT file containing its definition.

5.4.1   A Proposal for Solution

The algorithm for obtaining the entire set of Nash equilibria for a given game is based on two crucial points: the minimization ability of Fuzzy ASA and the use of an excellent seed finder. Here the second role is played by space-filling curves in the initial phase, based on deterministically sampling the objective function domain by means of an configurable discretization process. Hence, the minimization process starts by exploring the function domain, searching for promising starting points for the main minimization algorithm. The total amount of seeds is adjustable in the actual implementation and should be set proportionally to the dimension of the problem. It is also important to properly choose the number of function evaluations to be spent in the first, exploratory part—more dimensions ask for more function evaluations.

The general algorithm can be stated as:

- I—Set initial sampling parameters (discretization granularity and number of seeds to obtain);
- II—Obtain the seeds;
- III—Launch the main global minimization process using the current seed;
- IV—Examine whether a satisfactory number of Nash equilibria have been found;
- If not, advance to the next seed and re-run the process from the beginning (go to step III above);
- Emit the results and finish the session.

5.4.2   Definition of the Games to Be Tested

It is important to highlight that although finite strategic games seem somewhat "cold" or perhaps artificial, they really are able to model and truly represent practical problems, being useful in the corresponding solutions. Their most common applications are in the microeconomic area, modeling oligopolies and several other market situations. Therefore, when applying the theory, players represent competitors, payoff functions quantify individual gains in each possible configuration, and Nash equilibria represent conditions in which players could stay in a sustainable way.

One remark about notation is in order here: the presentation format to be used in the description of the payoff functions is hierarchical with matrix nesting, that is, more encompassing submatrices' positions indicate the leftmost indices, and the more internal ones complete the full indexing of a given association. For example, in the first payoff function of Problem 1 we have $u_1(1, 1, 1, 1) = 1.131$ and $u_1(2, 1, 2, 2) = 4.383$. Another example, with the full "translation" in terms of multidimensional matrices, is as follows:

The matrix

$$\begin{bmatrix} [9\ 0]\ [0\ 3] \\ [0\ 3]\ [9\ 0] \end{bmatrix}$$

means that

$$u(1, 1, 1) = 9, \ u(1, 1, 2) = 0$$
$$u(1, 2, 1) = 0, \ u(1, 2, 2) = 3$$
$$u(2, 1, 1) = 0, \ u(2, 1, 2) = 3$$
$$u(2, 2, 1) = 9, \ u(2, 2, 2) = 0.$$

Now it is time to describe the test examples.

Problem 1. This game features four players and two pure strategies per player. It presents three equilibria and the GAMBIT specification file is $2 \times 2 \times 2 \times 2$.nfg.

The version of GAMBIT used for the solution was able to find the following equilibria:

{ (1, 0, 1, 0, 1, 0, 0, 1), (0.1004, 0.8996, 0, 1, 0, 1, 0.2699, 0.7301), (0, 1, 1, 0, 0, 1, 1, 0) }

with four probability mass functions presented in each solution, corresponding to the four players. The payoff functions are specified below:

Player 1 (u_1)

$$\begin{bmatrix} \begin{bmatrix} 1.131\ 2.326 \\ 1.223\ 5.255 \end{bmatrix} \begin{bmatrix} 4.452\ 2.747 \\ 4.564\ 5.634 \end{bmatrix} \\ \begin{bmatrix} 4.225\ 1.478 \\ 4.483\ 4.383 \end{bmatrix} \begin{bmatrix} 7.566\ 1.759 \\ 7.247\ 4.642 \end{bmatrix} \end{bmatrix}$$

Player 2 (u_2)

$$\begin{bmatrix} \begin{bmatrix} 1.210\ 2.422 \\ 1.358\ 5.334 \end{bmatrix} \begin{bmatrix} 4.549\ 2.243 \\ 4.326\ 5.675 \end{bmatrix} \\ \begin{bmatrix} 5.277\ 2.544 \\ 5.764\ 1.436 \end{bmatrix} \begin{bmatrix} 4.655\ 2.705 \\ 4.943\ 1.897 \end{bmatrix} \end{bmatrix}$$

Player 3 (u_3)

$$\begin{bmatrix} \begin{bmatrix} 2.426\ 3.222 \\ 2.234\ 2.643 \end{bmatrix} \begin{bmatrix} 1.655\ 3.518 \\ 1.762\ 2.455 \end{bmatrix} \\ \begin{bmatrix} 4.837\ 1.973 \\ 4.995\ 4.864 \end{bmatrix} \begin{bmatrix} 7.076\ 1.735 \\ 7.362\ 4.042 \end{bmatrix} \end{bmatrix}$$

Player 4 (u_4)

$$\left[\begin{bmatrix} 2.429 \; 3.024 \\ 2.238 \; 2.873 \end{bmatrix} \begin{bmatrix} 1.347 \; 3.062 \\ 1.576 \; 6.523 \end{bmatrix}\right.$$
$$\left.\begin{bmatrix} 5.645 \; 7.486 \\ 5.754 \; 5.267 \end{bmatrix} \begin{bmatrix} 4.423 \; 4.043 \\ 4.382 \; 3.830 \end{bmatrix}\right]$$

Problem 2. In this case, we have a four person normal form game with two pure strategies per player. The game presents five mixed equilibria and the GAMBIT specification file is g3.nfg. The equilibria set found by using GAMBIT is given by:

{ (0.2, 0.8, 1, 0, 1, 0, 0.6667, 0.3333),
(1, 0, 1, 0, 0.4286, 0.5714, 0.8, 0.2),
(1, 0, 0.5643, 0.4357, 0.5318, 0.4682, 0.4255, 0.5745),
(0.6318, 0.3682, 1, 0, 0.6338, 0.3662, 0.5872, 0.4128),
(0.7111, 0.2889, 0.6938, 0.3062, 0.6201, 0.3799, 0.3646, 0.6354) }

with four probability mass functions in each one, corresponding to the four players. The payoff functions are defined by:

Player 1

$$\left[\begin{bmatrix} -3 \; -3 \\ -4 \; -2 \end{bmatrix} \begin{bmatrix} -3 \; -3 \\ -4 \; -1 \end{bmatrix}\right.$$
$$\left.\begin{bmatrix} -4 \; -1 \\ -4 \; -3 \end{bmatrix} \begin{bmatrix} -1 \; -1 \\ -6 \; -8 \end{bmatrix}\right]$$

Player 2

$$\left[\begin{bmatrix} -4 \; -6 \\ -2 \; -4 \end{bmatrix} \begin{bmatrix} -5 \; -2 \\ -7 \; -4 \end{bmatrix}\right.$$
$$\left.\begin{bmatrix} -5 \; -3 \\ -3 \; -6 \end{bmatrix} \begin{bmatrix} -8 \; -6 \\ -3 \; -5 \end{bmatrix}\right]$$

Player 3

$$\left[\begin{bmatrix} -1 \; -6 \\ -2 \; -2 \end{bmatrix} \begin{bmatrix} -3 \; -3 \\ -6 \; -5 \end{bmatrix}\right.$$
$$\left.\begin{bmatrix} -3 \; -3 \\ -4 \; -6 \end{bmatrix} \begin{bmatrix} -3 \; -2 \\ -1 \; -5 \end{bmatrix}\right]$$

Player 4

$$\left[\begin{bmatrix} -6 \; -2 \\ -3 \; -6 \end{bmatrix} \begin{bmatrix} -5 \; -5 \\ -1 \; -3 \end{bmatrix}\right.$$
$$\left.\begin{bmatrix} -3 \; -4 \\ -2 \; -7 \end{bmatrix} \begin{bmatrix} -3 \; -1 \\ -2 \; -5 \end{bmatrix}\right]$$

Problem 3. This is a 5 player game, with two pure strategies available to each player. The game is characterized by 5 Nash equilibria. The corresponding GAMBIT file is $2 \times 2 \times 2 \times 2 \times 2$.nfg and the equilibria set found is given by

{ (0.1441, 0.8559, 0.2584, 0.7416, 1, 0, 1, 0, 0, 1),
(0, 1, 0, 1, 1, 0, 0.7959, 0.2041, 0.5589, 0.4411),
(1, 0 , 0, 1, 0.1528, 0.8472, 0.6990, 0.3010, 1, 0),
(1, 0, 0, 1, 0, 1, 0.1185, 0.8815, 0.5564, 0.4436),
(1, 0, 0.2300, 0.7700, 0.6311, 0.3689, 0.6994, 0.3006, 1, 0) }

with five probability mass functions in each one, corresponding to the five players, and the payoff functions as below:

Player 1

$$
\begin{bmatrix}
\begin{bmatrix} [\ 7.247\ 4.943\ ] \ [\ 4.837\ 5.645\ ] \\ [\ 1.759\ 2.705\ ] \ [\ 3.518\ 3.062\ ] \end{bmatrix} \begin{bmatrix} [\ 1.223\ 1.358\ ] \ [\ 1.762\ 1.576\ ] \\ [\ 4.120\ 4.976\ ] \ [\ 4.864\ 5.267\ ] \end{bmatrix} \\
\begin{bmatrix} [\ 2.537\ 2.417\ ] \ [\ 2.355\ 5.347\ ] \\ [\ 2.778\ 1.646\ ] \ [\ 1.740\ 4.192\ ] \end{bmatrix} \begin{bmatrix} [\ 2.675\ 2.238\ ] \ [\ 2.426\ 2.429\ ] \\ [\ 2.747\ 2.243\ ] \ [\ 2.643\ 2.873\ ] \end{bmatrix}
\end{bmatrix}
$$

Player 2

$$
\begin{bmatrix}
\begin{bmatrix} [\ 7.362\ 4.382\ ] \ [\ 1.478\ 2.544\ ] \\ [\ 1.735\ 4.043\ ] \ [\ 1.131\ 1.210\ ] \end{bmatrix} \begin{bmatrix} [\ 2.234\ 2.238\ ] \ [\ 5.634\ 5.675\ ] \\ [\ 7.324\ 7.723\ ] \ [\ 1.428\ 1.521\ ] \end{bmatrix} \\
\begin{bmatrix} [\ 1.427\ 1.624\ ] \ [\ 2.798\ 1.403\ ] \\ [\ 2.342\ 1.64\ ] \ \ [\ 3.803\ 2.685\ ] \end{bmatrix} \begin{bmatrix} [\ 1.132\ 1.985\ ] \ [\ 2.326\ 2.422\ ] \\ [\ 3.518\ 3.062\ ] \ [\ 4.564\ 4.326\ ] \end{bmatrix}
\end{bmatrix}
$$

Player 3

$$
\begin{bmatrix}
\begin{bmatrix} [\ 4.642\ 1.897\ ] \ [\ 1.973\ 7.486\ ] \\ [\ 4.452\ 4.549\ ] \ [\ 2.426\ 2.429\ ] \end{bmatrix} \begin{bmatrix} [\ 5.255\ 5.334\ ] \ [\ 2.455\ 6.523\ ] \\ [\ 4.483\ 5.764\ ] \ [\ 4.248\ 4.533\ ] \end{bmatrix} \\
\begin{bmatrix} [\ 4.542\ 4.572\ ] \ \ [\ 2.832\ 1.38\ ] \\ [\ 2.164\ 5.349\ ] \ [\ 3.464\ 2.678\ ] \end{bmatrix} \begin{bmatrix} [\ 5.369\ 5.395\ ] \ [\ 3.222\ 3.024\ ] \\ [\ 1.223\ 1.358\ ] \ [\ 1.762\ 1.576\ ] \end{bmatrix}
\end{bmatrix}
$$

Player 4

$$
\begin{bmatrix}
\begin{bmatrix} [\ 4.042\ 3.83\ ] \ \ [\ 7.566\ 4.655\ ] \\ [\ 1.655\ 1.347\ ] \ [\ 2.326\ 2.422\ ] \end{bmatrix} \begin{bmatrix} [\ 2.643\ 2.873\ ] \ [\ 1.335\ 1.734\ ] \\ [\ 4.995\ 5.754\ ] \ [\ 5.231\ 5.458\ ] \end{bmatrix} \\
\begin{bmatrix} [\ 7.577\ 7.969\ ] \ [\ 3.974\ 2.793\ ] \\ [\ 2.243\ 5.056\ ] \ [\ 1.323\ 4.792\ ] \end{bmatrix} \begin{bmatrix} [\ 4.669\ 4.274\ ] \ [\ 4.452\ 4.549\ ] \\ [\ 2.234\ 2.238\ ] \ [\ 5.634\ 5.675\ ] \end{bmatrix}
\end{bmatrix}
$$

Player 5

$$
\begin{bmatrix}
\begin{bmatrix} [\,4.225\ 5.277\,]\ [\,7.076\ 4.423\,] \\ [\,2.747\ 2.243\,]\ [\,3.222\ 3.024\,] \end{bmatrix} & \begin{bmatrix} [\,4.564\ 4.326\,]\ [\,4.434\ 4.271\,] \\ [\,4.383\ 1.436\,]\ [\,4.486\ 4.882\,] \end{bmatrix} \\[3em]
\begin{bmatrix} [\,2.236\ 5.174\,]\ [\,3.575\ 2.269\,] \\ [1.473\ 4.58\,]\ \ \ [\,1.214\ 4.462\,] \end{bmatrix} & \begin{bmatrix} [\,1.131\ 1.21\,]\ \ [\,1.655\ 1.347\,] \\ [\,5.255\ 5.334\,]\ [\,2.455\ 6.523\,] \end{bmatrix}
\end{bmatrix}
$$

Problem 4. This is a normal form game with three players and two pure strategies available to each player. This game presents 9 Nash equilibria. However, the latest available graphical version of Gambit at the time of this writing (0.2007.12.04) was able to find only seven points, terminating abnormally. Consequently, we will list below only the seven ones found by Gambit. The two missing equilibria were found by the proposed method to be (0.4, 0.6, 0.5, 0.5, 1/3, 2/3) and (0.5, 0.5, 0.4, 0.6, 0.25, 0.75). The corresponding Gambit file is $2 \times 2 \times 2$.nfg.

The equilibria set found by using Gambit is given by

{ (1, 0, 1, 0, 1, 0) , (1, 0, 0, 1, 0, 1), (0, 1, 0, 1, 1, 0),
(0,1,1,0,0,1), (0,1, 0.25, 0.75, 0.3333, 0.6667),
(0.5, 0.5, 0.5, 0.5, 1, 0), (0.3333, 0.6667, 1,0, 0.25, 0.75) }

with three probability mass functions in each one. The payoff functions are given by:

Player 1

$$
\begin{bmatrix} [\,9\ 0\,]\ [\,0\ 3\,] \\ [\,0\ 3\,]\ [\,9\ 0\,] \end{bmatrix}
$$

Player 2

$$
\begin{bmatrix} [\,8\ 0\,]\ [\,0\ 4\,] \\ [\,0\ 4\,]\ [\,8\ 0\,] \end{bmatrix}
$$

Player 3

$$
\begin{bmatrix} [\,12\ 0\,]\ [\,0\ 6\,] \\ [\,0\ 6\,]\ [\,2\ 0\,] \end{bmatrix}
$$

Problem 5. This is a three agent coordination game with three strategies available to each player.The game presents 13 Nash equilibria.The GAMBIT file that corresponds to this game is coord333.nfg.

The equilibria set found by using GAMBIT is given by

{ (1,0,0,1,0,0,1,0,0), (0,0,1,0,0,1,0,0,1), (0,0,1,0,1,0,1,0,0), (0,0,1,1,0,0,0,1,0),
(0,1,0,0,0,1,1,0,0), (0,1,0,0,1,0,0,1,0), (0,1,0,1,0,0,0,0,1),
(1,0,0,0,1,0,0,0,1), (0.5,0.5,0,0.5,0.5,0,0.5,0.5,0), (0,0.5,0.5,0,0.5,0.5,0,0.5,0.5),

(0.5,0,0.5,0.5,0,0.5,0.5,0,0.5),
(0.3333,0.3333,0.3333,0.3333,0.3333,0.3333,0.3333,0.3333,0.3333),
(1,0,0,0,0,1,0,1,0) }

with three probability mass functions in each one, corresponding to the three players. The payoff functions are defined by:

First player

$$
\begin{bmatrix}
[1\ 0\ 0] & [0\ 0\ 0] & [0\ 0\ 0] \\
[0\ 0\ 0] & [0\ 1\ 0] & [0\ 0\ 0] \\
[0\ 0\ 0] & [0\ 0\ 0] & [0\ 0\ 1]
\end{bmatrix}
$$

Second player

$$
\begin{bmatrix}
[1\ 0\ 0] & [0\ 0\ 0] & [0\ 0\ 0] \\
[0\ 0\ 0] & [0\ 1\ 0] & [0\ 0\ 0] \\
[0\ 0\ 0] & [0\ 0\ 0] & [0\ 0\ 1]
\end{bmatrix}
$$

Third player

$$
\begin{bmatrix}
[1\ 0\ 0] & [0\ 0\ 0] & [0\ 0\ 0] \\
[0\ 0\ 0] & [0\ 1\ 0] & [0\ 0\ 0] \\
[0\ 0\ 0] & [0\ 0\ 0] & [0\ 0\ 1]
\end{bmatrix}
$$

Problem 6. This is a 2 player game with 4 strategies available to each player. This game has a total of 15 Nash equilibria. The corresponding Gambit file is coord4.nfg.

The equilibria set found by using Gambit is given by

{ (14/39, 14/39, 7/39, 4/39, 4/25, 6/25, 12/25, 3/25),
(0, 14/25, 7/25, 4/25, 0, 2/7, 4/7, 1/7),
(14/25, 0 , 7/25, 4/25, 4/19,0, 12/19, 3/19),
(0, 0, 7/11, 4/11, 0, 0, 4/5, 1/5),
(7/16, 7/16, 0, 1/8, 4/13, 6/13, 0, 3/13),
(0, 7/9, 0, 2/9, 0, 2/3, 0, 1/3), (7/9, 0, 0, 2/9, 4/7, 0, 0, 3/7),
(0,0,0,1, 0,0,0,1), (2/5,2/5,1/5,0, 2/11, 3/11, 6/11, 0), (0, 2/3,1/3,0, 0, 1/3, 2/3,0),
(2/3,0,1/3,0,1/4,0,3/4,0), (0,0,1,0,0,0,1,0),
(1/2,1/2,0,0,2/5,3/5,0,0), (0,1,0,0,0,1,0,0),
(1,0,0,0,1,0,0,0) }

with two probability mass functions in each one. The payoff functions are as follows:

First player

$$\begin{bmatrix} 3 & 0 & 0 & 0 \\ 0 & 2 & 0 & 0 \\ 0 & 0 & 1 & 0 \\ 0 & 0 & 0 & 4 \end{bmatrix}$$

Second player

$$\begin{bmatrix} 2 & 0 & 0 & 0 \\ 0 & 2 & 0 & 0 \\ 0 & 0 & 4 & 0 \\ 0 & 0 & 0 & 7 \end{bmatrix}$$

In the following the results obtained in [3] will be discussed.

5.5 Experimental Results

Previous work [16] showed that many optimization methods may be combined to multistart and deflection techniques in order to find simultaneously the Nash equilibria for the six games described above, except for the fifth problem, whose 13 Nash equilibria could not be found during a single run.

Furthermore, experimental data was presented, evidencing that not every execution of a single algorithm was capable to find all equilibria of all tests. In addition, the performance evaluation of CMA-ES, DE and PSO in each test case compared the results of each single algorithm to the number of different Nash equilibria detected, being also based on the mean number of function evaluations required to compute different Nash equilibria. They also investigated the differences caused by the choice between deflection and multistart techniques. The tests consisted of 30 numerical experiments for each of the algorithms, and the stopping criterion was to reach objective function (5.7) values less than 0.00000001.

In [3] the same experimental conditions were kept, and the experiments using multistart Fuzzy ASA showed that the method is able to find all Nash equilibria of all six tests in all thirty runs, within the established numerical tolerance. So, the results were very stable, leading to the conclusion that multistart Fuzzy ASA was effective.

According to [3] all tests used 5000 function evaluations for each dimension of the domain during the exploratory phase, and 15 additional starting points for each dimension of the problem. For more information about ASA configuration, please, see [4].

By the way, in [3] important information about certain ASA parameters is cited. For example, if cost and parameter quenching are enabled, certain runtime variables (User_Quench_Cost_Scale[] and User_Quench_Param_Scale[]) may be used to change the parameter and cost scales of temperature schedules. Those are the ones on which Fuzzy ASA acts in order to modify the dynamics of the global optimization process. Although the default is to assign the annealing value of 1 to all elements that might be defined otherwise, if values greater than 1 are set, quenching is activated. For the sake of monitoring the optimization process and record the best parameters found so far, the runtime variables Best_Cost and Best_Parameters[] are used too.

5.6 Conclusion

In this chapter it was demonstrated that it is possible, for a general purpose meta-heuristic optimization method, to find all Nash equilibria of a given strategic game in an automatic way. The presented paradigm does not use particular features of any game in order to improve its performance and, even so, it is able to find all equilibrium points of each game in the test set. The main goal is to show the effectiveness of the Fuzzy ASA method with respect to its ability to find all Nash equilibria of finite normal form games and, to that end, the optimization formulation of the problem of computing Nash equilibria is adopted.

It is noteworthy the fact that the detection of multiple Nash equilibria in single executions of the algorithm was possible thanks to the preprocessing phase, during which space-filling curves were used to sample the objective function landscape.

References

1. Aguiar e Oliveira, Jr, H., Petraglia, A.: Global optimization using space-filling curves and measure-preserving transformations. In: Gaspar-Cunha A., et al. (eds.) Soft Computing in Industrial Applications, pp. 121–130. Springer, Berlin (2011)
2. Aguiar e Oliveira Jr, H., Petraglia, A.: Global optimization using dimensional jumping and fuzzy adaptive simulated annealing. Appl. Soft Comput. 11, 4175–4182 (2011)
3. Aguiar e Oliveira Jr, H., Petraglia, A.: Establishing Nash equilibria of strategic games: a multistart fuzzy adaptive simulated annealing approach. Appl. Soft Comput. 19, 188–197 (2014)
4. Aguiar e Oliveira Jr, H., Ingber, L., Petraglia, A., Petraglia, M.R., Machado, M.A.S.: tochastic Global Optimization and Its Applications with Fuzzy Adaptive Simulated Annealing. Springer, Berlin (2012)
5. Boryczka, U., Juszczuk, P.: Approximate Nash equilibria in bimatrix games. Comput. Collect. Intell. Technol. Appl., Lect. Notes Comput. Sci. 6923, 485–494 (2011)
6. Boryczka, U., Juszczuk, P.: A new evolutionary approach for computing Nash equilibria in bimatrix games with known support. Cent. Eur. J. Comput. Sci. 2(2), 128–142 (2012)
7. Boryczka, U., Juszczuk, P.: Differential evolution as a new method of computing Nash equilibria. Trans. Comput. Collect. Intell. IX, Lect. Notes Comput. Sci. 7770, 192–216 (2013)
8. Ingber, L.: Adaptive Simulated Annealing (ASA): lessons learned. Control Cybern. 25(1), 33–54 (1996)
9. Koh, A.: Locating multiple Nash equilibria in hierarchical games. Online Conf. Soft Comput. Ind. Appl.—WSC
10. Koh, A.: An evolutionary algorithm based on Nash dominance for equilibrium problems with equilibrium constraints. Appl. Soft Comput. 12(1), 161–173 (2012)

11. Mallozzi, L.: An application of optimization theory to the study of equilibria for games: a survey. Cent. Eur. J. Op. Res. **21**(3), 523–539 (2013)
12. McKelvey, R.D.: A Liapunov Function for Nash Equilibria. Technical Report. California Institute of Technology, Pasadena (1991)
13. McKelvey, R.D., McLennan, A.: Computation of equilibria in finite games. In: Amman, H.M., Kendrick, D.A., Rust, J. (eds.) Handbook of Computational Economics, Handbooks in Economics (13), 1, pp. 87–142. North-Holland, Amsterdam (1996)
14. McKelvey, R.D., McLennan, A., Turocy, T.L.: Gambit: Software Tools for Game Theory. http://www.gambit-project.org, (2013)
15. Osborne, M.J., Rubinstein, A.: A Course in Game Theory. MIT Press, Cambridge (1994)
16. Pavlidis, N.G., Parsopoulos, K.E., Vrahatis, M.N.: Computing Nash equilibria through computational intelligence methods. J. Comput. Appl. Math. **175**, 113–136 (2005)
17. Sagan, H.: Space-Filling Curves. Springer, New York (1994)

Chapter 6
Generalized Nash Equilibrium Problems and Fuzzy ASA

Abstract As an extension of the standard Nash equilibrium concept, the generalized Nash equilibrium (GNE) makes it possible to model and solve more general problems in several scenarios. Its most prominent advantage resides in that the GNE concept allows objective functions and constraints associated to each player to depend on the strategies of other agents, creating a more realistic environment. By studying GNE properties, problems in several fields, including Engineering and Economics, may be modelled and solved in an easier way. In this chapter a solution algorithm based on the Fuzzy ASA algorithm is introduced, evidencing that it is possible to transform many complex tasks into constrained global optimization problems—as such, they can be solved, in principle, by any effective global optimization algorithm, but here the main tool is Fuzzy ASA. The intention is to show that the presented approach may offer a simpler alternative for solving this type of problem in a less limited way, that is, not imposing strong conditions on the defining functions. After the theoretical explanation, many examples are presented in order to demonstrate the efficacy of the method.

6.1 Introduction

This chapter exposes an alternative approach for solving Generalized Nash Equilibrium Problems (GNEP's, for short), considering they have been used actively in many areas of knowledge during the recent past and research activity on this topic is increasing. The presented method transforms the original problem into a constrained optimization one and, by means of global optimization algorithms, searches for points that represent the desired generalized Nash equilibria for the original task. Accordingly, the efficacy of the method depends primarily on the quality of the optimization algorithm used in a given task. Considering that GNEP's are used to solve many problems in different branches of science and technology, and researchers in different fields work independently, many results are not yet widely known, although some fundamental ideas were well-established, such as the problem definition and certain approaches considered adequate for solving the problem, as using Nikaidô–Isoda functions, for example. The object of study itself (GNEP) has been referred to

© Springer International Publishing Switzerland 2016   93
H. Aguiar e Oliveira Junior, *Evolutionary Global Optimization,*
Manifolds and Applications, Studies in Systems, Decision and Control 43,
DOI 10.1007/978-3-319-26467-7_6

as a social equilibrium problem, coupled constraint equilibrium problem, or abstract economy. Here the expression **generalized Nash equilibrium problem** is adopted and precisely defined in the sequel [3].

In GNEP's the cost functions assigned to each player and the corresponding strategies are real valued and continuous. Besides, they allow each players' constraints to depend on their opponents' strategies. This characteristic extends the corresponding one in standard Nash equilibrium problems, in which players' actions are independent of the strategies chosen by others. Therefore, the coupled constrained action space in GNEP's makes them more difficult to solve than conventional problems, and algorithms able to solve them can also be applied to standard NEP's.

GNEP's are defined in an environment consisting of N players, where player i controls n_i variables, stored into a vector $x^i \in \mathbb{R}^{n_i}$. Naming $\mathbf{x}$ the vector formed by all these variables, the result is

$$\mathbf{x} \triangleq \begin{pmatrix} x^1 \\ x^2 \\ \vdots \\ x^N \end{pmatrix} \tag{6.1}$$

with $\mathbf{x} \in \mathbb{R}^{\mathbf{n}}$ and $n = n_1 + n_2 + \cdots + n_N$. We also denote by $\mathbf{x}^{-i}$ the vector $\mathbf{x}$ without the x^i sub-vector.

$$\mathbf{x}^{-i} \triangleq \begin{pmatrix} x^1 \\ x^2 \\ \vdots \\ x^{i-1} \\ x^{i+1} \\ \vdots \\ x^n \end{pmatrix} \tag{6.2}$$

It is also common practice to write $\mathbf{x} = (x^i, \mathbf{x}^{-i})$, observing that the notation $(x^i, \mathbf{x}^{-i})$ does not mean that the components of $\mathbf{x}$ have been reordered, and x_i is the first block. A further condition is that all players are assumed to make their decisions at the same time (simultaneous strategy choices), and each one is assigned one loss or cost function $\phi_i(\mathbf{x})$ ($\phi_i : \mathbb{R}^n \to \mathbb{R}$). Therefore, when overall decision vector is $\mathbf{x}$, player i has to lose the value $\phi_i(\mathbf{x})$. So, the cost function of a specific player may depend on other players' decision vectors, the same being true with respect to the strategy sets of each agent—this does not happen in standard Nash equilibrium problems.

Considering players as rational agents, that is, assuming that, given the decision vector $\mathbf{x}^{-i}$ of other players, they will try (individually) to choose a decision vector x^i that minimizes their cost function, it is expected that a search for the solution of the optimization problem below occurs at each game instance

$$\min_{x^i} \ \phi_i(x^i, \mathbf{x}^{-i}) \quad s.t. \quad x^i \in X_i(\mathbf{x}^{-i}) \tag{6.3}$$

with $X_i(\mathbf{x}^{-i}) \overset{\triangle}{=} \{x^i \in \mathbb{R}^{n_i} | (x^i, \mathbf{x}^{-i}) \in X\}$, and X not empty, closed and convex subset of $\mathbb{R}^n$. X is the global strategy set.

Hence, it is possible to define a vector $\mathbf{x}^* = (x^{*,1}, x^{*,2}, \ldots, x^{*,N})^T$ as being a generalized Nash equilibrium if, for all $i = 1, 2, \ldots, N$, the subvector $x^{*,i}$ solves the following optimization problem

$$\min_{x^i} \ \phi_i(x^i, \mathbf{x}^{*,-i}) \quad s.t. \quad x^i \in X_i(\mathbf{x}^{*,-i}) \tag{6.4}$$

or

$$\phi_i(x^{*,i}, \mathbf{x}^{*,-i}) \leq \phi_i(x^i, \mathbf{x}^{*,-i}) \quad \forall x^i \in X_i(\mathbf{x}^{*,-i}) \tag{6.5}$$

for $i = 1, \ldots, N$.

GNEP's have other identifications, as social equilibrium problems, or Nash equilibrium problems with shared constraints, among others. They were introduced by Debreu [2] in 1952, and ever since have been predominantly used in the mathematical economics field—only recently its importance outside economics has finally been recognized. Typical applications can be found in telecommunications and electrical engineering, computer science, market regulation, or ecological settings.It is important to mention that GNEP's often possess multiple solutions and, in general, different solutions of certain equilibrium problems may have different implications, differently from a typical optimization problem in which all optimal solutions can be regarded as having equal value. From this viewpoint it may be meaningful to choose a particular generalized equilibrium, featuring a certain additional property. In [20] it is proposed the solution concept known as "normalized equilibrium", that is a special GNE characterized by conditions imposed on Lagrange multipliers associated with the constraints in each player's problem. The uniqueness of a normalized equilibrium can be established under certain theoretical conditions [20, 23].

6.2 An Overview of Current Results

Currently it is known that standard NEP's may be converted into variational inequality problems [5]. Furthermore, it was established that it is possible to handle GNEP's as quasi-variational inequalities (QVI's) [1, 14] but, on the other hand, considering that efficient methods for solving QVI's are rare, this type of characterization might not be very useful, in practical terms. In addition, it was shown [7, 8] that certain solutions of the GNEP may be found by solving a certain standard VIP associated to the given GNEP.

Also, some VIP based methods demand a higher degree of smoothness of the objective functions when compared to other approaches, based on Nikaidô–Isoda functions [19], which play a central role in the related literature. By the way, relaxation methods using Nikaidô–Isoda functions are studied in [17, 21], and a proximal-like method based on them is introduced in [8].

Reference [22] reports the use of a regularized version of the Nikaidô–Isoda function in the synthesis of transformed optimization problems whose global minima are the normalized solutions of the original GNEP—the function and its regularized version are precisely defined, obtaining a constrained optimization problem that is equivalent to the GNEP. However, the cost function relative to this optimization problem is not smooth. Hence a smooth optimization problem is obtained, whose solutions characterize the class of normalized Nash equilibria of the GNEP. Finally, it is shown how the previous techniques can be used in order to get a smooth unconstrained optimization reformulation of the normalized GNEP solutions.

The regularized Nikaidô–Isoda function was employed earlier for solving standard NEP's in [13]. The standard NEP is known to be a special case of an equilibrium programming problem [9], and this fact is emphasized in [22] by interpreting GNEP's as particular instances of such a type of problem.

Even though GNEP's are important paradigms in terms of modeling power, current approaches aiming at their solution are somewhat dispersed, when thinking of the diversity of theoretical tools used when trying to solve them. Rosen studied jointly convex GNEP's [20] and this scenario was very frequent in the literature until recently. He presented the first algorithm for solving such GNEP's and other methods were later developed for these problems, like the relaxation method [17, 21], based again on Nikaidô–Isoda functions. As expected, other approaches may be employed when solving jointly convex problems, and among them it is possible to find the recently proposed methods from [22–24], which are also based on the Nikaidô–Isoda function and consequently computationally intensive. There are also the variational inequality approaches [7], which are also limited to the jointly convex case. In the case of unrestricted GNEP's things get even harder, considering that it is known that a GNEP can be reduced to a quasi-variational inequality [1, 3, 14]. It occurs that the development of globally convergent algorithms for the solution of QVI's is still a difficult task, and this transformation is of little practical use, even using gap functions to change GNEP's into optimization problems. Nikaidô–Isoda functions are a great tool when transforming GNEP's into optimization problems, but the computational load is expressive and the conditions for establishing convergence are reasonably complex. Evaluating the whole scenario we can state that, excepting penalty methods and algorithms for very specific applications, the study of solutions for general GNEP's in more general conditions is still in its infancy. Penalty algorithms aimed at solving GNEP's are based on the well-known penalization approach, by eliminating the hard inter-related constraints in a GNEP and reducing it to a less complex NEP. To solve this latter problem, it is possible to apply several algorithms. The penalization approach to GNEP's is relatively recent and was initiated by Fukushima and Pang in [12], in which it is proposed a sequential penalty approach, and an infinite sequence of differentiable penalized problems is solved.

The techniques of using exact penalty approaches in which a single nondifferentiable NEP is solved was studied in [6], being also addressed in [11], where some conditions are stated under which a penalty approach can be employed to find solutions for GNEP's. Even though some problems have to be addressed, as a complete understanding of theoretical conditions under which useful results can be established for a penalty approach, penalty based methods are very promising approaches for solving GNEP's [4].

In [16] a metaheuristic algorithm is used in order to study the Nash normalized equilibrium, which can be obtained by transforming the GNEP into a bi-level program with an optimal value of zero in the upper level. It is proposed a differential evolution based bi-level programming algorithm that uses stochastic ranking to handle constraints and solve the resulting formulation—in this work Nikaidô–Isoda functions play an important role as well.

Reference [22] describes a well-known approach, cited previously in this chapter and presented here for the sake of illustrating the computational complexity usually coupled to methods aimed at solving GNEP's. In that work, the main theoretical tool used in obtaining one of the optimization reformulations of the GNEP is the Nikaidô–Isoda function, defined by

$$\Psi(\mathbf{x}, \mathbf{y}) \stackrel{\Delta}{=} \sum_{i=1}^{N} [\phi_i(x^i, \mathbf{x}^{-i}) - \phi_i(y^i, \mathbf{x}^{-i})] \tag{6.6}$$

From this, it is possible to define

$$V(\mathbf{x}) \stackrel{\Delta}{=} \sup_{y \in \Omega(\mathbf{x})} \Psi(\mathbf{x}, \mathbf{y}), \quad \mathbf{x} \in X \tag{6.7}$$

supposing that the supremum always exists for some $\mathbf{y} \in \Omega(\mathbf{x})$.

From this it is possible to conclude that $V(\mathbf{x})$ is nonnegative for all $\mathbf{x} \in \Omega(\mathbf{x})$, and that $\mathbf{x}^*$ is a solution of the GNEP if and only if $\mathbf{x}^* \in \Omega(\mathbf{x}^*)$ and $V(\mathbf{x}^*) = 0$. Hence, finding solutions for the GNEP is equivalent to computing global minima of the corresponding optimization problem

$$\min_{\mathbf{x} \in \Omega(\mathbf{x})} V(\mathbf{x}) \tag{6.8}$$

having

$$\Omega(\mathbf{x}) \stackrel{\Delta}{=} X_1(\mathbf{x}^{-1}) \times \cdots \times X_N(\mathbf{x}^{-N}) \tag{6.9}$$

Notice that this problem usually has a nontrivial feasible set, considering that $\Omega(\mathbf{x})$ depends on the overall vector $\mathbf{x}$. However, remembering a result established in [22] (Lemma 2.1), problem (6.8) is found to be equivalent to the following one

$$\min_{\mathbf{x} \in X} V(\mathbf{x}) \tag{6.10}$$

Observe that to each evaluation of $V(\mathbf{x})$ corresponds a complete optimization cycle, defined by (6.7).

Even though Nikaidô–Isoda functions are already consolidated in the literature, they have some not very favorable features. For example, for a given vector $\mathbf{x}$, the

supremum in (6.7) may not exist except if additional assumptions (like the compactness of X) are assured, and besides, the supremum is often not attained at an
isolated, single point, implying that the mapping V and, therefore, the corresponding
optimization reformulation (6.10), may be nondifferentiable. To solve these problems, a simple regularization of the Nikaidô–Isoda function may be used (please, see
[22])—this idea was used earlier in other contexts [10, 13, 18]. So, it is reasonable to
conclude that many present approaches to the GNEP are computationally expensive,
and a new approach capable of simplifying the search process and reduce the overall
processing cost would be really useful.

6.3 The Basic Algorithm

6.3.1 Fundamental Facts

As said earlier, this chapter aims to show that it is feasible to solve GNEP's after
a single optimization run. In principle, the general idea can be applied by means
of any metaheuristic global optimization method and, as usual, we chose the Fuzzy
Adaptive Simulated Annealing paradigm. The effort is focused on demonstrating
that the Fuzzy ASA method may solve GNEP's, faced as constrained optimization
problems, without the use multi-level optimization techniques. Initially it is necessary
to change the problem of computing generalized Nash equilibria, that can be faced as
a set of coupled and simultaneous constrained optimization problems, into a single
one, whose solutions are exactly the optimizers of the original problem. This step
is often done by means of penalty techniques, which are frequently effective, that
is, make it possible to solve the original problem. In addition, and considering the
large applicability of Fuzzy ASA, the presented method is likely to be more flexible
than some of the already existing ones in many cases, given its ability to deal with
nondifferentiable and nonconvex functions defined on nonconvex domains. To make
things more precise, we establish the target GNEP.

Determine $\mathbf{x}^* = (x^{*,1}, x^{*,2}, \ldots, x^{*,N})^T \in \mathbb{R}^n$ such that, for $i = 1, 2, \ldots, N$, $x^{*,i}$
is a solution for the problem

$$\min_{x^i} \ \phi_i(x^i, \mathbf{x}^{*,-i}) \quad s.t. \quad x^i \in X_i(\mathbf{x}^{*,-i}) \tag{6.11}$$

or

$$\phi_i(x^{*,i}, \mathbf{x}^{*,-i}) \le \phi_i(x^i, \mathbf{x}^{*,-i}) \quad \forall x^i \in X_i(\mathbf{x}^{*,-i}) \tag{6.12}$$

with $i = 1, \ldots, N$.

Being $X_i(\mathbf{x}^{*,-i}) \stackrel{\Delta}{=} \{x^i \in \mathbb{R}^{n_i} | (x^i, \mathbf{x}^{*,-i}) \in X\}$. X, the global strategy set, is supposed to be a nonempty and closed subset of $\mathbb{R}^n$, and $(n = \sum_{i=1}^{N} n_i)$.

Here, the functions ϕ_i are continuous and not necessarily differentiable or convex, being bounded on compact domains—in the literature, more restrictive conditions are usually present.

6.3.2 The Algorithm

- **Preparation**

  For each ϕ_i and $X_i(\mathbf{x}^{-i})$, find the conditions that can give rise to numerical constraints which, after inserted into penalized objective functions, may guide the algorithm toward adequate solutions.

  So, whenever loss functions are differentiable it is possible to force partial gradients to zero, in order to help in the search for equilibria, taking into account that this is a necessary condition in this specific situation.

  The same reasoning applies to the sets $X_i(\mathbf{x}^{-i})$, particularly whenever they are expressed as parametric inequalities, as below

  $$X_i(\mathbf{x}^{-i}) \stackrel{\Delta}{=} \{x^i \in \mathbb{R}^{n_i} \mid g^i(x^i, \mathbf{x}^{-i}) \le 0\} \tag{6.13}$$

- **Step 1**

  For each potential condition obtained in the preparation stage, create an additive term (to be added to the global penalty function) capable of converting the corresponding imposition into a numerical value, so as to make lower values correspond to more favorable regions of the search domain.

- **Step 2**

  Compose the cost function by determining weights that will be used in the global optimization step. As usual, these weights will multiply each penalty term, dictating the relative importance of each one during the optimization process.

  The final cost function has the following aspect

  $$C(\mathbf{x}) \stackrel{\Delta}{=} \sum_{i=1}^{N} F_{\phi_i} \times \phi_i(\mathbf{x}) + \sum_{i=1}^{N} F_{Grad_i} \times ||\nabla^{n_i} \phi_i(\mathbf{x})||^2$$

  $$+ \sum_{i=1}^{N} F_{Constr_i} \times Constr_i(\mathbf{x}) + \sum_{i=1}^{N_{Attr}} F_{Attr_i} \times Attr_{\phi_i}(\mathbf{x}) \tag{6.14}$$

in which

$F_{\phi_i}, F_{Grad_i}, F_{Constr_i}$ and F_{Attr_i} correspond to the weights,
$\nabla^{n_i} \phi_i(\mathbf{x})$ is the partial gradient of $\phi_i(\mathbf{x})$, if defined,

$Constr_i(\mathbf{x})$ is the penalty term corresponding to $\phi_i(\mathbf{x})$,

$Attr_{\phi_i}(\mathbf{x})$ corresponds to the ith auxiliary numerical function, aimed at driving the optimization process into adequate regions (N_{Attr} occurrences).

- **Step 3**

  Start the chosen global minimization algorithm (Fuzzy ASA), using $C(\mathbf{x})$ as cost function.

- **Decision Point**

  Collect all subvectors from the overall $\mathbf{x}$ found in the previous step and evaluate the results. If they are not adequate, go to step 2, re-designing specific penalty terms, otherwise emit the subvectors as the final solution.

6.3.3 Practical Facts

Although any effective global optimization algorithm could be used to optimize the global cost function, the choice of Fuzzy ASA was motivated by the fact that it has shown to be very effective in many different types of problems, and able to deal with really hard optimization tasks.

6.4 Numerical Simulations

Now it is time to evaluate the quality of the exposed ideas, presenting some examples of GNEP's and corresponding solutions. In this fashion it will be possible to demonstrate the efficacy of the method described in this chapter and offer the reader a general idea of how to apply it in practice.

6.4.1 2-Dimensional Example

Here we discuss an example taken from [24], and originally suggested by Rosen [20]. It consists of two cost functions and one global set, determining the joint constraints. The loss functions are

$$\theta_1(x_1, x_2) \overset{\Delta}{=} 0.5x_1^2 - x_1 x_2 \tag{6.15}$$

$$\theta_2(x_1, x_2) \overset{\Delta}{=} x_2^2 + x_1 x_2 \tag{6.16}$$

and the joint constraints are given by

$$X \overset{\Delta}{=} \{x \in \mathbb{R}^2 \mid x_1 \geq 0, \ x_2 \geq 0, \ x_1 + x_2 \geq 1\} \tag{6.17}$$

The objective is to find at least one Nash equilibrium point (x_1^*, x_2^*).
We have

$$\frac{\partial \theta_1}{\partial x_1}(x_1, x_2) = x_1 - x_2 \tag{6.18}$$

$$\frac{\partial \theta_2}{\partial x_2}(x_1, x_2) = 2x_2 + x_1 \tag{6.19}$$

The penalty function is

$$P(x_1, x_2) \stackrel{\Delta}{=} max\{1 - (x_1 + x_2), 0\} \tag{6.20}$$

and the penalized cost function becomes

$$C(x_1, x_2) \stackrel{\Delta}{=} \theta_1(x_1, x_2) + \theta_2(x_1, x_2) + \alpha_P \times P(x_1, x_2) \tag{6.21}$$

It is easy to infer that x_1 is the control variable of player 1 and x_2 is the control variable of player 2. The conditions $x_1 \geq 0$ and $x_2 \geq 0$ do not need to be translated into penalty terms because the ASA system allows us to control the point generation process so as to keep candidates inside a given hyper-rectangle. Here it is possible to incorporate the constraints in the specification very early so as to simplify the resulting loss function and the corresponding solution process. In addition, the partial derivatives of cost functions in the penalty terms were not used, considering that they only vanish at the same time outside the feasible region, and this might "confuse" the minimization algorithm. After submitting $C(x_1, x_2)$ to the Fuzzy ASA algorithm, it was found a minimizer that almost coincides with the unique normalized Nash equilibrium $(1, 0)$ reported in [24] for this problem—it is easy to verify that this point is a GNE for this example.

As expected, the partial derivatives of the cost functions (θ_1 and θ_2) at this point are almost unitary—as supposed, they are far from zero. The values of the cost functions at the above equilibrium point are near 0.500 and 0, respectively. Also, in the experiments the fixed value used for α_P (weight for the only constraint present in the overall penalized cost function) was set in 10E+08, and candidate points generated by ASA were located inside the rectangle $[0, 10] \times [0, 10]$. The total number of objective function evaluations spent was 3,636,755 on average, after 30 simulations.

6.4.2 N-Dimensional Example

The present example was taken from [4] (Example A.1) and presents a variant of an internet switching model introduced by Kesselman et al. [15]. The simulation environment consists of N players, each of which featuring only one control variable $x^i \in \mathbb{R}$. The individual objective functions are given by

$$\theta_i(\mathbf{x}) \triangleq \frac{-x^i}{\sum_{i=1}^{N} x^i} \left(1 - \frac{\sum_{i=1}^{N} x^i}{B} \right), \quad i = 1, \ldots, N \tag{6.22}$$

being B a constant.

The first player is bound to the interval $[0.3, 0.5]$, and the other ones obey the conditions

$$\sum_{i=1}^{N} x^i \leq B \tag{6.23}$$

$$x^i \geq 0.01, \quad i = 2, \ldots, N \tag{6.24}$$

Anticipating a possible local minimum condition for some loss functions, it is advisable to obtain the expression for the corresponding partial derivatives. The one relative to the cost function of player i is given by

$$\frac{\partial \theta_i}{\partial x^i}(\mathbf{x}) = \frac{x^i}{\left(\sum_{i=1}^{N} x^i\right)^2} - \frac{1}{\sum_{i=1}^{N} x^i} + \frac{1}{B} \tag{6.25}$$

Setting $N = 10$ and $B = 1$, the reported solution for this problem in [4] is given by the following vector of $\mathbb{R}^{10}$

(0.29923815223336, 0.06951127617805, 0.06951127617805,
0.06951127617805, 0.06951127617805, 0.06951127617805,
0.06951127617805, 0.06951127617805, 0.06951127617805,
0.06951127617805).

It occurs that the value of the first component (0.29923815223336) disagrees with the constraint $0.3 \leq x^1 \leq 0.5$. On the other hand, the algorithm presented here was able to find the point

(0.3000172773, 0.0697282309484, 0.069580539,
0.069426857095, 0.069426856858, 0.069426856949,
0.069426856306, 0.069426856966, 0.069426856944,
0.069426856947).

that satisfies all constraints and features very low values for partial derivatives, apart from the first variable, because of its distinguished constraint.

It is worth noting that convergence was achieved by means of standard ASA (without activation of the fuzzy controller).

The final penalized cost function is

$$C(\mathbf{x}) \overset{\Delta}{=} \sum_{i=1}^{N} \theta_i(\mathbf{x}) + P_{Der} \times \sum_{i=2}^{N} \frac{\partial \theta_i}{\partial x^i}(\mathbf{x}) + P_{Constr} \times max \left\{ \sum_{i=1}^{N} x^i - B, 0 \right\} \quad (6.26)$$

in which P_{Der} and P_{Constr} are constants.

Similarly to the previous experiment, constraints $x_i \geq 0.01$ ($i \geq 2$) and $x_1 \in [0.3, 0.5]$ are not included in the penalty terms, considering that the ASA system allows us to set the point generation process so that candidate points are created inside a hyper-rectangle. It is important to reinforce that partial derivatives corresponding to the first player were not used, considering that the corresponding cost function has additional restrictions on its domain and it is not clear whether some unconstrained minimum is present in the feasible region. According to the obtained results, the partial derivatives of the players' cost functions, $\theta_i(i \geq 2)$, are practically null, indicating that the individual functions reached their unconstrained minima, except θ_1. The fixed value used for P_{Der} is 200,000,000, the one for P_{Constr} is 20,000, and the candidate points in ASA were generated inside the hyper-rectangle $[0.3, 0.5] \times [0.01, 1] \times \cdots \times [0.01, 1]$. The total number of objective function evaluations needed to reach the final result was 1,135,666, on average, after executing 30 test runs.

6.4.3 N-Dimensional Example

The present test appears in [4] (Example A.2) and is similar to the previous one. As above, the scenario is composed of N players, each one possessing one control variable $x^i \in \mathbb{R}$. Individual cost functions are the same as before, apart from agents 2, 3, 4, and 5, with expressions given by

$$\theta_i(\mathbf{x}) \overset{\Delta}{=} \frac{-x^i}{\sum_{i=1}^{N} x^i} \left(1 - \frac{\sum_{i=1}^{N} x^i}{B} \right)^2, \quad i = 2, 3, 4, 5 \quad (6.27)$$

in which B = constant.

Additional constraints with the form $0.99 \leq \sum_{i=1}^{N} x^i$ are imposed to agents 5 and 6, and upper bounds for variables linked to players 9 and 10 are set at 0.06 and 0.05, respectively.

Partial derivatives corresponding to the cost functions of players 2, 3, 4, and 5 are

$$\frac{\partial \theta_i}{\partial x^i}(\mathbf{x}) = \frac{x^i - \sum_{i=1}^{N} x^i}{\left(\sum_{i=1}^{N} x^i \right)^2} \left(1 - \frac{\sum_{i=1}^{N} x^i}{B} \right)^2 + \frac{2x^i}{B \times \sum_{i=1}^{N} x^i} \left(1 - \frac{\sum_{i=1}^{N} x^i}{B} \right) \quad (6.28)$$

Putting $N = 10$ and $B = 1$, the solutions for this problem are reported (in [4]) to be the following

(0.29962894677774, 0.00997828224734, 0.00997828224734, 0.00997828224734, 0.59852469355630, 0.02187270661760, 0.00999093169361, 0.00999093169361, 0.00999093169361, 0.00999093169361)

and

(0.29962898846513, 0.00997828313762, 0.00997828313762, 0.00997828313762, 0.59745624992082, 0.02220301920403, 0.01013441012117, 0.01013441012117, 0.01013441012117, 0.01013441012117)

Similarly to a previous test, the values of first components slightly violate the constraint $0.3 \le x^1 \le 0.5$ for player 1.

Proceeding as usual, the algorithm presented earlier was able to produce the point

(0.5, 0.08776934, 0.0878395, 0.087727456, 0.0878394, 0.01, 0.01, 0.01, 0.06, 0.05).

The values obtained for the respective partial derivatives at this point are a good approximation for

(0.5, 0, 0, 0, 0, 0, 0, 0, 0, 0).

Again, the convergence was achieved by means of standard ASA.

The definitive cost function used in this example takes the form

$$
C(\mathbf{x}) \triangleq \sum_{i=1}^{N} \theta_i(\mathbf{x}) + P_{Der} \times \sum_{i=2}^{8} \frac{\partial \theta_i}{\partial x^i}(\mathbf{x})
$$

$$
+ P_{Constr} \times \left[\left(max\left\{ \sum_{i=1}^{N} x^i - B, 0 \right\} \right)^2 + \left(max\left\{ 0.99 - \sum_{i=1}^{N} x^i, 0 \right\} \right)^2 \right]
$$

$$
\tag{6.29}
$$

in which P_{Der} and P_{Constr} are fixed weighting factors.

Once more some conditions are not present in the penalty terms because the ASA system provides mechanisms which allow the user to tune the point generation process so as to generate candidates inside a chosen hyper-rectangle. Note that, again, partial derivatives corresponding to players 1, 9 and 10 were not used in the making of the final penalized cost function, considering that their specific cost functions have severe restrictions on their domains, what may provoke certain difficulties in assuring the occurrence of null derivatives inside the corresponding domains. In the tests, the value used for P_{Der} was 11,000,000, and P_{Constr} was assigned the value 1,000,000.

Besides, ASA generated candidate points inside the hyper-rectangle $[0.3, 0.5] \times [0.01, 1.0] \times \cdots \times [0.01, 1.0] \times [0.01, 0.06] \times [0.01, 0.05]$. The total number of objective function evaluations needed to reach the final result was 2,886,325, on average, after performing 30 sessions.

6.4.4 Another 2-Dimensional Example

This problem was also taken from [4] (Example A.12). There are $N = 2$ players, each of which with one control variable $x^i \in \mathbb{R}$, corresponding to a duopoly model. Their isolated cost functions have similar expressions, namely,

$$\theta_i(\mathbf{x}) \triangleq x^i (\rho(x^1 + x^2) + \lambda - d), \quad i = 1, 2 \tag{6.30}$$

with $x^i \in [-10, 10]$.

Expressions for partial derivatives are

$$\frac{\partial \theta_1}{\partial x^1}(\mathbf{x}) = \rho(2x^1 + x^2) + \lambda - d \tag{6.31}$$

$$\frac{\partial \theta_2}{\partial x^2}(\mathbf{x}) = \rho(x^1 + 2x^2) + \lambda - d \tag{6.32}$$

Putting $d = 20, \lambda = 4, \rho = 1$, in [4] it is reported the approximated solution $(5.33331555561568, 5.33331555561568)^T$, and the exact solution $(\frac{16}{3}, \frac{16}{3})^T$.

By using the algorithm presented in this chapter it was possible to find the minimizer
$(5.3333333333333, 5.33333333333333)$, with derivatives
$(0.000000000000009, 0.0000000000000000)$ and minimum values
-28.4444444444444 and -28.4444444444444, respectively.

As above, the convergence was achieved by means of standard ASA.

The penalized cost function used here is given by

$$C(\mathbf{x}) \triangleq P_{Fun} \times \sum_{i=1}^{2} \theta_i(\mathbf{x}) + P_{Der} \times \sum_{i=1}^{2} \left| \frac{\partial \theta_i}{\partial x^i}(\mathbf{x}) \right| \tag{6.33}$$

in which P_{Der} and P_{Fun} are weighting constants.

Here, P_{Der} is 1,100,000 and P_{Fun} is 1,500, and the candidate points and ASA generated pints inside the rectangle $[-10, 10] \times [-10, 10]$. On average, 79,557 objective function evaluations were sufficient to reach the final result, after 30 runs.

6.5   Conclusion

This chapter presented a method and corresponding results that evidenced the feasibility of solving GNEP's in a single optimization step by means of a general purpose, metaheuristic optimization method. One important intention is to demonstrate the effectiveness of the Fuzzy ASA method with respect to its ability to find solutions for GNEP's. To get there, the original optimization formulation of the task (which is to calculate generalized Nash equilibria) was transformed. Hence, the central problem, previously faced as a set of coupled and simultaneous constrained optimization problems, was changed into a single one, by means of a penalty approach. Such a transformation was successful and, in addition, given the nature of the chosen global optimization method (Fuzzy ASA), the presented method can be more general than some of the already existing ones in many circumstances, taking into account that it is able to deal with nondifferentiable functions, defined on nonconvex domains. Not only the presented method but also this new way of facing GNEP's are certainly significant contributions to the field, having in mind that multi-stage optimization phases are eliminated and the whole process gets simplified. As the presented results show, the method is effective.

References

1. Bensoussan, A.: Points de Nash dans le cas de fonctionelles quadratiques et jeux differentiels lineaires a N personnes. SIAM J. Control **12**, 460–499 (1974)
2. Debreu, G.: A social equilibrium existence theorem. Proc. Natl. Acad. Sci. **38**, 886–893 (1952)
3. Facchinei, F., Kanzow, C.: Generalized Nash Equilibrium Problems, Preprint 290. University of Würzburg, Würzburg, Germany, Institute of Mathematics (2009)
4. Facchinei, F., Kanzow, C.: Penalty methods for the solution of generalized Nash equilibrium problems (with complete test problems). Technical Report, Institute of Mathematics, University of Würzburg, Würzburg, Germany (2009)
5. Facchinei, F., Pang, J.-S.: Finite-Dimensional Variational Inequalities and Complementarity Problems, vol. I. Springer, New York (2003)
6. Facchinei, F., Pang, J.-S.: Exact penalty functions for generalized Nash problems. In: Di Pillo, G., Roma, M. (eds.) Large Scale Nonlinear Optimization, pp. 115–126. Springer, New York (2006)
7. Facchinei, F., Fischer, A., Piccialli, V.: On generalized Nash games and VIs. Technical Report, Department of Computer and System Sciences "A. Ruberti", Università di Roma "La Sapienza", Rome, Italy (2006)
8. Flåm, S.D., Ruszczyński, A.: Noncooperative convex games, computing equilibria by partial regularization. IIASA Working Paper 94-42, Laxenburg, Austria (1994)
9. Flåm, S.D., Antipin, A.S.: Equilibrium programming using proximal-like algorithms. Math. Program. **78**, 29–41 (1997)
10. Fukushima, M.: Equivalent differentiable optimization problems and descent methods for asymmetric variational inequality problems. Math. Program. **53**, 99–110 (1992)
11. Fukushima, M.: Restricted generalized Nash equilibria and controlled penalty algorithm. Technical Report 2008–007. Kyoto University, Department of Applied Mathematics and Physics(2008)

12. Fukushima, M., Pang, J.-S.: Quasi-variational inequalities, generalized Nash equilibria and multi-leader-follower games. Comput. Manage. Sci. **2**, 21–56 (2005)

13. Gürkan, G., Pang, J.-S.: Approximations of Nash equilibria. Technical Report,Department of Mathematical Sciences, Rensselaer Polytechnic Institute, Troy, NY (2006)

14. Harker, P.T.: Generalized Nash games and quasivariational inequalities. Eur. J. Oper. Res. **54**, 81–94 (1991)

15. Kesselman, A., Leonardi, S., Bonifaci, V.: Game-theoretic analysis of internet switching with selfish users. Internet Netw. Econ. **3828**, 236–245 (2005)

16. Koh, A.: Differential evolution based bi-level programming algorithm for computing normalized Nash equilibrium. In: Gaspar-Cunha, A., et al. (eds.) Soft Computing in Industrial Applications, AISC 96. Springer, Berlin (2011)

17. Krawczyk, J.B., Uryasev, S.: Relaxation algorithms to find Nash equilibria with economic applications. Environ. Model. Assess. **5**, 63–73 (2000)

18. Mastroeni, G.: Gap functions for equilibrium problems. J. Glob. Opt. **27**, 411–426 (2003)

19. Nikaidô, H., Isoda, K.: Note on noncooperative convex games. Pac. J. Math. **5**, 807–815 (1955)

20. Rosen, J.B.: Existence and uniqueness of equilibrium points for concave N-person games. Econometrica **33**, 520–534 (1965)

21. Uryasev, S., Rubinstein, R.Y.: On relaxation algorithms in computation of noncooperative equilibria. IEEE Trans. Autom. Control **39**, 1263–1267 (1994)

22. von Heusinger, A., Kanzow, C.: Optimization reformulations of the generalized Nash equilibrium problem using Nikaido-Isoda type functions, Preprint 269. University of Würzburg, Würzburg, Germany, Institute of Mathematics (2006)

23. von Heusinger, A., Kanzow, C., Fukushima, M.: Newton's method for computing a normalized equilibrium in the generalized Nash game through fixed point formulation, Preprint 286. University of Würzburg, Würzburg, Germany, Institute of Mathematics (2009)

24. von Heusinger, A., Kanzow, C.: SC^1 optimization reformulations of the generalized Nash equilibrium problem. Opt. Methods Softw. **23**, 953–973 (2008)

Chapter 7
Studying Coalitions

Abstract A deep study assessing the feasibility of coalition formation in electric energy auctions is presented. A stochastic global optimization algorithm, when applied to the calculation of Nash–Cournot equilibria in several scenarios, makes it possible to obtain quantitative results concerning the profitability of coalition formation processes in diverse environments. Auxiliary Nash equilibrium problems are solved by transforming the original problem into a global optimization one and constructing cost functions which translate the associated constraints into mathematical relations, reflecting the benefit maximization trend of typical energy conversion and transmission firms. It is also indicated how to use the algorithm to estimate coupled constraint equilibria occurring when restrictions are imposed to businesses or marketplaces. In addition, the suggested method computes players' payoffs in many configurations, comparing their profits and production levels under different market elasticities. Furthermore, solutions are based on cooperative game theory concepts, such as the bilateral Shapley value. It is shown that the adequacy of creating certain coalition configurations depends critically on demand × price elasticity relationships. A case study based on the IEEE 30-bus system is used, for the sake of presenting and discussing in detail the paradigm. The presented method is far-reaching and uses the solution of generalized Nash equilibrium problems to obtain numerical data that will take us to the final decisions. As seen in the previous chapter, generalized Nash equilibrium problems address extensions of the well-known standard Nash equilibrium concept, making it possible to model and study more general configurations. As said before, GNEP's have a larger scope, considering that they allow both objective functions and constraints of each player to depend on the strategies of other players. As can be observed from the literature, the study of such problems finds endless applications in several areas, including Medicine, Engineering, and Management Science, for example.

7.1 Introduction

Recently some countries reformed their electric energy markets, changing conventional, vertically integrated monopolies into deregulated agents, oftentimes oligopolies. Hence, competition among firms has increasingly been fostered by these

© Springer International Publishing Switzerland 2016   109
H. Aguiar e Oliveira Junior, *Evolutionary Global Optimization,*
Manifolds and Applications, Studies in Systems, Decision and Control 43,
DOI 10.1007/978-3-319-26467-7_7

transformed marketplaces, in which buyers and sellers may trade electricity in auctions or by means of conventional transactons. Profit maximization strategies used by electric utilities in the past are being replaced by more effective bidding methods. Usually, the aim of electric utilities is to maximize profit in their transactions, where prices are determined by suppliers, consumers, transmission line operators, and other players. In this type of market, perfect competition is very hard to attain in practice, due to the reduced number of competing agents. In another dimension, network limitations may affect market competitiveness, since bidders can produce bottlenecks, leading to a reasonable raise in unitary prices [3]. So, it seems reasonable to suppose that we are facing an imperfect market.

Accordingly, it is expected that oligopolistic models be used to investigate the dynamics of electricity markets, mainly as a consequence of the restructuring trend that started some years ago. The modeling paradigm presented in this chapter is similar to those in [3, 4], taking into account that both papers deal with Nash–Cournot equilibria in electricity markets, although there are some differences between them. In contrast with the method in [4], that depends on the existence of a solution to a system of equations and inequalities, resulting from mixed complementarity (KKT) conditions, the one in [3] relies on deterministic function minimization procedures. In addition, an important difference is in the type of equilibrium that each study is trying to determine: in [4] the aim is to find a Nash–Cournot equilibrium that could also satisfy a market clearing condition. From a different perspective, here and in [3] the search is for coupled constraint equilibria, a relatively new solution concept in game theory, in which the strategy space is jointly restricted for all agents. Hence, the two lines of reasoning adopt the latter kind of equilibrium as an adequate solution concept for many significant electricity market related games. Nikaidô–Isoda functions and relaxation algorithms are composed in [5, 11] in order to create an algorithm aimed at solving infinite games. This method is interesting because the most complex procedure is to minimize a particular multivariate function. Also, a sequential improvement of the Nikaidô–Isoda function is obtained by means of a relaxation algorithm that is shown to converge to a Nash equilibrium for a expressive class of problems, admitting nondifferential payoffs and coupled constraint games [9, 11]. The possibility of solving games with constrained strategy spaces is important when modeling electric energy markets, considering that in many problems related to electricity generation and distribution the strategy spaces of players is coupled. It occurs, for example, because of the capacity constraints and Kirchhoff's laws (inescapable physical conditions), that is, there are joint impositions on the combined strategy space of all players.

Consequently, the subset of strategies that a certain agent may use depends on the other agents' decisions and, in simultaneous games, conventional noncooperative game theory concepts are unable to handle satisfactorily the problem. Fortunately, Rosen's normalized equilibrium, idealized to solve games subject to coupled constraint sets (coupled constraint games), can be applied [9]. Presently, this kind of problem is also addressed by underlying techniques associated to generalized Nash equilibrium problems, and the main point of this chapter is to study in detail the process of coalition formation in important scenarios—here it is done by using acase study presented in [3], in the context of electricity markets. The approach here is to handle the task by solving several GNEP's, viewed as global optimization problems, with the help of the Fuzzy ASA algorithm—the method exposed in this chapter follows closely the material presented in [1].

7.2 Establishing the Problem

In Ref. [3], two significant problems are presented—the first one considers an electricity market that uses the IEEE 30-bus system [7], and was chosen as the object of investigation in this chapter. When facing that problem, it is assumed that there are three generating companies, each of which possessing a small number of generating units, as displayed in Table 7.1. P^g and P^C are the power generation of a unit and a company, respectively. With the same notation, it is assumed therein that the generation cost of unit i is given by the expression

$$C_i(P_{gi}) = (c_i/2)P_{gi}^2 + d_i P_{gi} + e_i \qquad (7.1)$$

whose coefficients are shown in Table 7.2.

Supposing that energy demand decreases with rising market prices, the generic format for demand functions corresponding to a typical day may be expressed as $P_{load}(p) = P_{load}^0(p) + ap$, in which $P_{load}^0(p)$ is the total power demand level expected for a pre-established time interval, and a is the elasticity of the demand with respect to the price. Following this line, the expression linking price and overall production of the IEEE 30-bus system in a given interval is assumed in [3] to be

$$P_{load}(p) = 189.2 - 0.5p \qquad (7.2)$$

It is clear that $a = -0.5$ and $P_{load}^0(p) = 189.2$.
Transforming the relationship into

$$p = 378.4 - 2P_{load}(p) \qquad (7.3)$$

with

Table 7.1 Limits for power generation

Company	Generator	P_{min}^g (MW)	P_{max}^g (MW)	P_{min}^C (MW)	P_{max}^C (MW)
1	1	0	80	0	80
2	2	0	80	0	130
	3	0	50		
3	4	0	55	0	125
	5	0	30		
	6	0	40		

Table 7.2 Cost coefficients
for generating units

Generator	c_i ($\$/MW^2h$)	d_i ($\$/MWh$)	e_i ($\$/h$)
1	0.04	2	0
2	0.035	1.75	0
3	0.125	1	0
4	0.0166	3.25	0
5	0.05	3	0
6	0.05	3	0

$$P_{load}(p) = \sum_{i=1}^{ng} P_{gi} - P_{loss} \qquad (7.4)$$

Here ng = total number of generators and P_{loss} = transmission losses throughout the
system.

Although energy losses are not being taken into account in this chapter or in [3],
they may be easily inserted into the model without interfering with the presented
method. This is possible due to the flexibility of the paradigm, considering that such
a insertion would cause only a change in the form of cost functions.

Now it is possible to obtain the crucial functions associated to each agent. The
profit ϕ_j relative to firm j, possessing ng_j generating units, is found to be

$$\phi_j = \left(378.4 - 2\sum_{i=1}^{ng} P_{gi}\right) \sum_{k=S(j)+1}^{S(j)+ng_j} P_{gk} - \sum_{k=S(j)+1}^{S(j)+ng_j} ((c_k/2)P_{gk}^2 + d_k P_{gk} + e_k) \qquad (7.5)$$

with variables corresponding to produced power subject to $P_{min}^{gj} \le P_{gj} \le P_{max}^{gj}$, as
displayed in Table 7.1. Besides, terms $S(j)$ make the expressions more compact, and
have values $S(1) = 0$, $S(2) = 1$, and $S(3) = 3$.

Reference [3] reports the use of a software package to solve the adopted case. However, the same results could have been obtained with the traditional Nash–Cournot
equilibrium conditions translated into a system of equations. Although the results
were obtained based on the assumption that the three companies do not cooperate
with each other, some players may consider the creation of a consortium, aiming
at increasing their profits. In other words, final production levels of firms could be
lower than the ones resulting from the shown results, in case they decide to cooperate. Contexts in which the number of involved agents is small open the possibiliy of
studying all possibilities of coalition formation. Hence, by enumerating all possible
groupings and creating several games where the players are not necessarily firms,
but coalitions, the presented algorithm can be applied to each one, computing Nash
equilibria relative to the each particular configuration. Therefore, cooperation among
agents creates many different scenarios, with different Nash equilibria. However, the
final gains of individual agents is not found by the main algorithm, needing additional
processing in another direction.

So, ideas from cooperative game theory, such as the bilateral Shapley value (BSV) [8] and the kernel [2] can furnish adequate ways to understand how each collective profit (derived from Nash equilibrium values) can be divided among the firms in each coalition. These methods represent feasible ways to handle coalitions and allocate profits after they are formed. In practice, the quest for profit optimization can lead companies to join others and form a new player, composed of two or more agents.

In the chosen example [3], configurations corresponding to all possible coalition combinations were studied, and the values found express the profit obtained by each particular coalition, composed of agents combined in different configurations. The cited values can be obtained by finding Nash equilibria in previously cited games, and in previous work [3] values are calculated by applying a relaxation algorithm to the Nikaidô–Isoda function corresponding to each setting, and then obtaining the final individual profits per coalition after the convergence of the respective algorithm. The adopted approach at first finds all possible subsets of individual players, creating several coalition structures—after that, it faces each coalition as an individual entity, in order to make it possible to find stable profit and production levels for each case. Hence, in order to determine the coalitions that are actually formed, it is necessary to compute the values corresponding to each game, noting that they always correspond to the minimal profits that a coalition can guarantee for itself against any other coalition.

For certain elasticity values, it may be possible that no equilibrium values exist, and consequently individual firm profit assigments show to be unfeasible, at least using this kind of approach. On the oher hand, for more elastic demand regions, bilateral Shapley value solutions may be shown to exist. Figure 7.1 illustrates what occurs in terms of elasticity when pivoting around the price intercept. Observe that, for different elasticity coefficients, games can become subadditive, superadditive, or none of them because, depending on the coalition values or joint profits, it may be better to stay alone or join other players.

Please, note that the experiments were done by fixing the intercept with the unitary price axis, varying only the slope of the price-demand line by varying elasticity values. In this fashion, to smaller elasticity values correspond slower speeds in the decrease of market prices when overall demand increases. Below, some important characteristics of the Shapley value are described. For more details, the interested reader may consult [8].

The Shapley value is one solution concept for transferable utility games, assigning a unique payoff distribution for the grand coalition to every such type of game. It attributes to each player his average marginal contribution to the game. In other words, its computation is based on averaging the values of contributions of each player in all possible "entering" orders. So, denoting a transferable utility game by (N, v), the Shapley value payoff associated to player $i \in N$ is

$$\Phi_i(N, v) = \sum_{S \subseteq N \,:\, i \notin S} \frac{|S|!(n - |S| - 1)!}{n!} [v(S \cup \{i\}) - v(S)] \qquad (7.6)$$

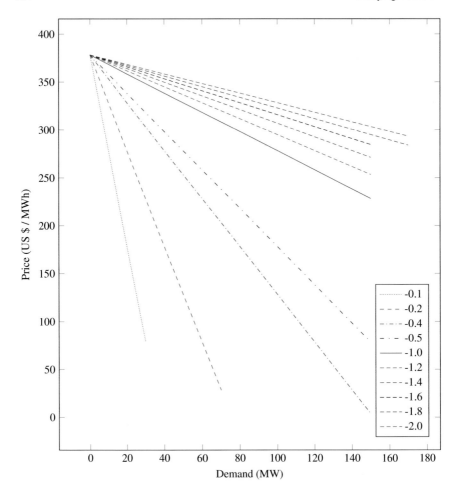

Fig. 7.1 Demand x price curves in several scenarios

Its definition by the assignment to each player of the average marginal contribution may be considered a justification for this solution concept by itself. In the literature there exist some axiomatic characterizations for this topic, being this kind of approach reasonably frequent in cooperative game theory. It is worth citing an axiomatic characterization of the Shapley value that can be found in [10]. For bigger games it may be more interesting to work with the previous formula than to use the definition based on marginal vectors.

The aim in this chapter is to investigate in detail the elasticity influence on the formation of coalitions, at the same time extending the study in [3] and offering ample numerical data about this significant subject. Hence, many results are shown, representing final agreements on splitting profits if acting collectively. Furthermore, it is not hard to see that price-demand elasticity is the main factor when trying to increase the chance of cooperation among oligopolistic firms. This is so because with a more elastic behavior, higher prices will be paid for a given demand level, and the amount to share will be larger. This effect will make a final agreement more probable than when the total profit is smaller, and all players tend to cluster in a grand coalition, with improved individual profits.

7.3 Simulations

Here, results obtained after simulating the adopted model and solving the corresponding GNEP's are presented. It is important to highlight that payoff functions found in the current application must have their signs inverted when using the techniques explained in the previous chapter—this is so because the algorithm assumes that we are dealing with cost (or loss) functions and aims to minimize the global objective function. As the target here is to maximize profits, this is equivalent to minimizing their additive opposites, as usual.

The simulation in this study uses the elasticity values in the set $\{-0.1, -0.2, -0.3, -0.4, -0.5, -1.0, -1.2, -1.3, -1.4\}$—in some specific cases, values in $\{-1.5, -1.6, -1.7, -1.8\}$ appear as well. Such a spectrum was assumed to be sufficient to reveal the most important qualitative aspects of this model's behavior, relatively to predicting whenever coalition formation is profitable.

When defining the problem in Sect. 7.2, it was stated that it features three companies (denoted by 1, 2, and 3) that can form five different coalition configurations, that is, $\{ \{1\}, \{2\}, \{3\} \}$, $\{ \{1, 2\}, \{3\} \}$, $\{ \{1, 3\}, \{2\} \}$, $\{ \{2, 3\}, \{1\} \}$, and $\{ \{1, 2, 3\} \}$. Excluding the last possibility, in each of them there exists a competitive scenario that can be handled by means of the Nash–Cournot equilibrium concept. In addition, it is necessary to address the issue of how to distribute the profits among the various firms participating in non-unitary agents, once we have the solutions corresponding to each coalition member.

Now it is time to discuss the results about production levels, profits, and unitary prices relative to each possible coalition structure—they are presented and illustrated with tables and graphs (Tables 7.3, 7.4, 7.5, 7.6 and 7.7, Figs. 7.2, 7.3, 7.4, 7.5, 7.6, 7.7, 7.8, 7.9, 7.10, 7.11, 7.12, 7.13, 7.14, 7.15 and 7.16). Production level measurement unit is MW, profits are in US dollars/h and prices in US dollars/MWh, as in [3].

7.3.1 Coalition { {1}, {2}, {3} }

When firms compete isolatedly, the simulations demonstrate that for elasticity values inferior to -0.5, production and profit levels exhibit a change in their evolution. It is possible to delimit three different parts in the corresponding graphs (Tables 7.3, 7.4, 7.5, 7.6 and 7.7):

- In the right part, after -0.5, production and profit figures behave in a similar manner for the three players, with unitary prices rising very slowly, almost linearly. It is clear that in this region all firms are able to supply additional energy demand, if necessary.

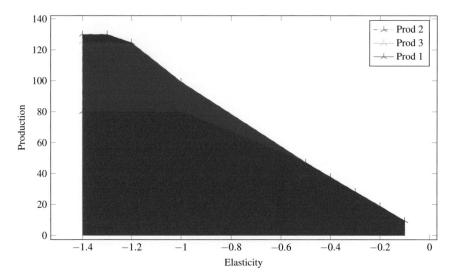

Fig. 7.2 Production—coalition { {1}, {2}, {3} }

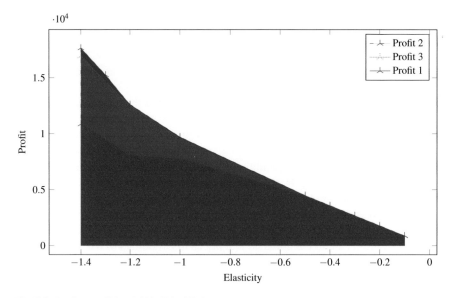

Fig. 7.3 Profit—coalition { {1}, {2}, {3} }

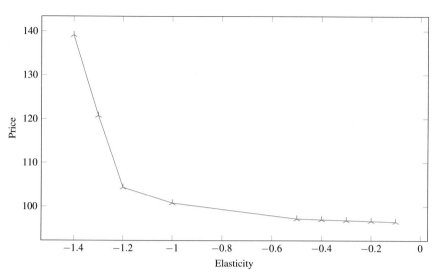

Fig. 7.4 Prices—coalition { {1}, {2}, {3} }

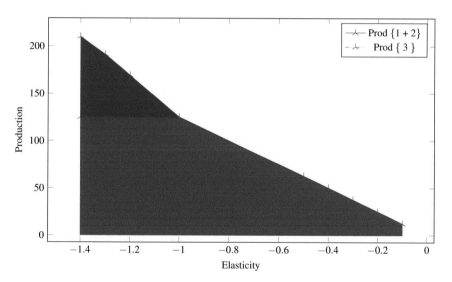

Fig. 7.5 Production—coalition { {1, 2}, {3} }

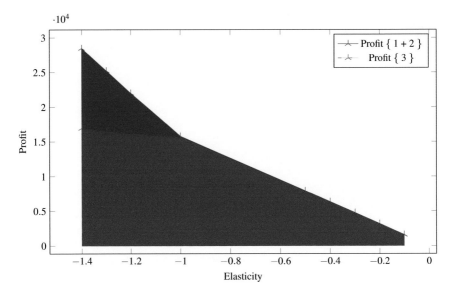

Fig. 7.6 Profit—coalition { {1, 2}, {3} }

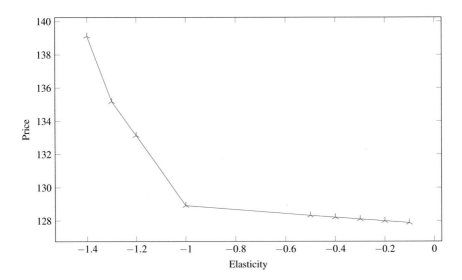

Fig. 7.7 Prices—coalition { {1, 2}, {3} }

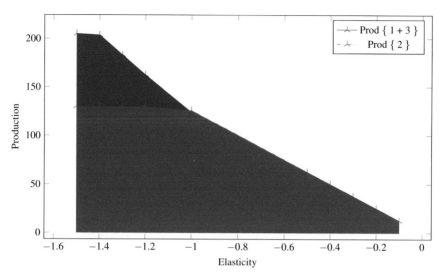

Fig. 7.8 Production—structure { {1, 3}, {2} }

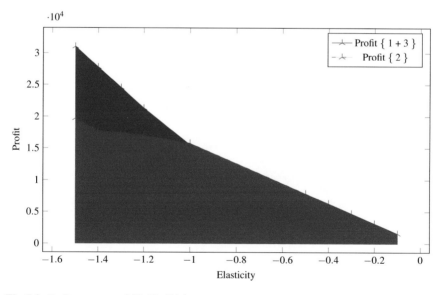

Fig. 7.9 Profit—structure { {1, 3}, {2} }

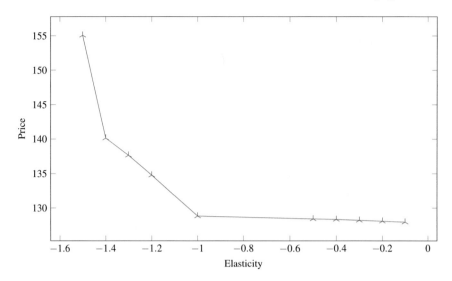

Fig. 7.10 Prices—structure { {1, 3}, {2} }

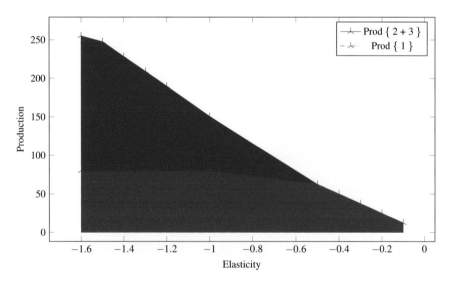

Fig. 7.11 Production—structure { {2, 3}, {1} }

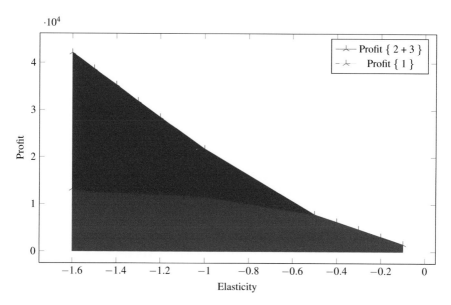

Fig. 7.12   Profit—structure { {2, 3}, {1} }

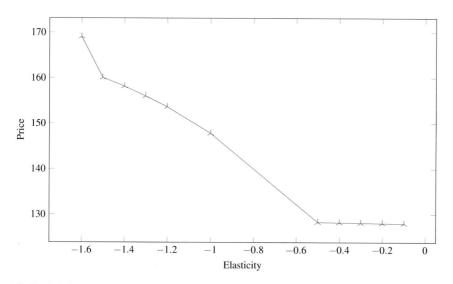

Fig. 7.13   Prices—structure { {2, 3}, {1} }

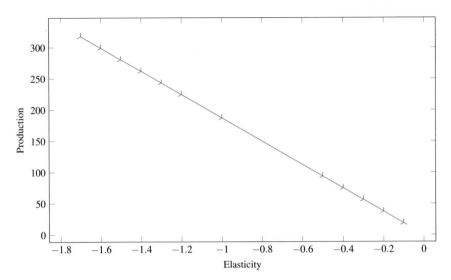

Fig. 7.14 Production—structure { {1, 2, 3} }

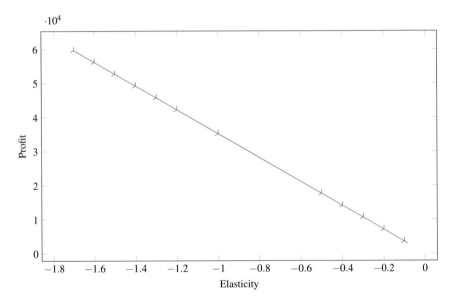

Fig. 7.15 Profit—structure { { 1, 2, 3 } }

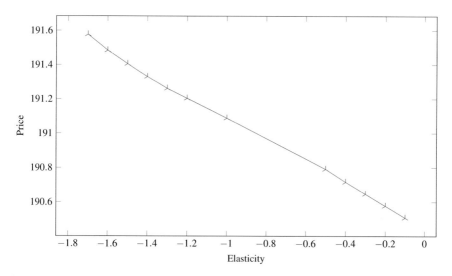

Fig. 7.16 Prices—structure { { 1, 2, 3 } }

- The central part displays players 2 and 3 bifurcating, in terms of production, with player 1 becoming unable to furnish further energy, considering that its production capacity was reached. As a result, unitary prices start to go up faster, especially for more negative values of elasticity.
- For elasticity values inferior to -1.2, players 2 and 3 hit their maximum production levels as well and unitary prices rise reasonably faster, evidencing the exhaustion in terms of production capacity and, as expected, all firms experience higher profits.

7.3.2 Coalition { {1, 2}, {3} }

This setting features a different type of evolution, showing a significant change for elasticity values inferior to -1.0—in this region individual production and profit levels behave differently, being possible to distinguish two regions in the graphs:

- On the right of -1.0, production figures and corresponding profits are similar for both agents, with unitary prices rising very slowly—in this situation all players are able to provide additional energy.
- Below -1.0 player {3} exhausts its production capacity and unitary prices rise very fast, as a result of the partial limitation in production ability. Consequently, players get larger profits until reaching the global production limit.

Table 7.3 IEEE 30-bus system Nash equilibrium results—{ {1}, {2}, {3} }

Elasticity	−0.1	−0.2	−0.3	−0.4	−0.5	−1.0	−1.2	−1.3	−1.4
Prod {1}	9.4076031735	18.778844542	28.111108919	37.405180499	46.661239642	80	80	80	80
Profit {1}	886.80233315	1770.2866376	2649.9391979	3525.886822	4398.1423129	7771.5834171	8059.7949963	9368.6153846	10841.142857
Prod {2}	9.4643	18.9085	28.3402	37.7565	47.1570	99.2866	124.7070	130	130
Profit {2}	898.32736694	1794.3101247	2689.9845073	3585.1460758	4479.7555918	9708.0691166	12591.228932	15233.75	17626.607143
Prod {3}	9.3229	18.6635	28.02	37.3943	46.7868	98.3686	124.1561	125	125
Profit {3}	870.09474209	1743.9562475	2621.7083309	3503.6020273	4389.6521393	9552.6550592	12483.212809	14612.104038	16912.928214
Un. Price	96.45258	96.645841	96.828841	97.010096	97.190093	100.744793	104.347437	120.707692	139.114286

Table 7.4 IEEE 30-bus system Nash equlibrium results—{ {1, 2}, {3} }

Elasticity	−0.1	−0.2	−0.3	−0.4	−0.5	−1.0	−1.2	−1.3	−1.4
Prod. {1, 2}	12.592049781	25.164922983	37.718977922	50.254198861	62.770672471	125.12678755	169.32617502	191.19157529	210
Profit {1, 2}	1589.8503275	3174.5449621	4756.9945454	6337.266957	7915.3675912	15786.644018	21987.324503	25212.458464	28467.75
Prod. {3}	12.460027649	24.917559857	37.371994239	49.823370361	62.271682602	124.36207988	125	125	125
Profit {3}	1554.100288	3108.2961522	4663.2852442	6219.0724562	7775.6472287	15558.478493	16164.665935	16420.606414	16912.928214
Un. price	127.879226	127.987586	128.096759	128.206077	128.31529	128.911133	133.128187	135.175711	139.14286

Table 7.5 IEEE 30-bus system Nash equilibrium results—{ {1, 3}, {2} }

Elasticity	-0.1	-0.2	-0.3	-0.4	-0.5	-1.0	-1.2	-1.3	-1.4	-1.5
Prod. {1, 3}	12.550309868	25.086246985	37.606708412	50.112028912	62.542257081	123.80623585	162.35901906	182.97808606	203.57642791	205
Profit {1, 3}	1577.1986751	3148.2124408	4718.2526799	6287.2440228	7846.9603656	15514.247382	21301.727058	24526.956335	27770.397987	31024.309167
Prod. {2}	12.493737961	24.975586716	37.447004170	49.90759693	62.448618948	125.7601621	130	130	130	130
Profit {2}	1577.864654	3152.4154067	4724.5394644	6294.2077849	7868.9387592	15773.499989	17061.522935	17435.941394	17758.79598	19700.416667
Un. price	127.959522	128.090831	128.220958	128.350935	128.418248	128.833602	134.767484	137.647626	140.131123	155.066667

Table 7.6 IEEE 30-bus system Nash equilibrium results—{ {2, 3}, {1} }

Elasticity	−0.1	−0.2	−0.3	−0.4	−0.5	−1.0	−1.2	−1.3	−1.4	−1.5	−1.6
Prod. (2, 3)	12.54108128	25.052242981	37.533632038	49.985368881	62.405170472	150.62130018	189.69205843	209.11697451	228.36380691	247.55753152	255
Profit (2, 3)	1583.9347259	3161.2464273	4732.3865644	6297.4037244	7856.6233977	21743.284872	28482.110065	31889.200885	35327.645897	38732.102472	42166.7675
Prod. [1]	12.499414763	25.013585489	37.542265549	50.085217925	62.643213197	80	80	80	80	80	80
Profit [1]	1571.7395502	3140.9705863	4707.6560402	6271.7625757	7833.5568879	11534.295985	12004.529438	12192.186184	12363.211033	12514.264986	13234
Un. price	127.995040	128.070858	128.147008	128.223533	128.303233	147.7787	153.656618	156.002327	158.140138	160.028312	169.025

Table 7.7 IEEE 30-bus system Nash equlibrium results—{{1 , 2 , 3 }}

Elasticity	-0.1	-0.2	-0.3	-0.4	-0.5	-1.0	-1.2	-1.3	-1.4	-1.5	-1.6	-1.7	-1.8
Prod. {1, 2, 3}	18.789152455	37.564517352	56.325911176	75.073390508	93.804211413	187.31170356	224.63566458	243.28040714	261.89790253	280.49251228	299.06543296	317.59937376	335
Profit (1, 2, 3)	3545.6771335	7084.2982574	10616.719433	14143.307554	17665.023826	35235.052585	42247.86796	45751.038214	49251.784872	52749.181748	56242.440158	59729.30726	63194.170278
Un. price	190.508475	190.577413	190.646963	190.716524	190.791577	191.088296	191.203613	191.261225	191.33007	191.404992	191.484104	191.576839	192.288889

7.3.3 Coalition { {1, 3}, {2} }

Although the qualitative behavior in this coalition configuration is similar to the previous one, numerical values differ, mainly with respect to the production limits of each agent. So, for values of elasticty greater than −1.0, production and corresponding profits behave in the same way for both agents, and unitary prices advance slowly—all players may furnish extra energy. On the other hand, below −1.0 player {2} becomes unable to furnish further energy and unitary prices rise faster, because of the lower production capacity. Hence, players get larger profits until reaching the global production limit.

7.3.4 Coalition { {2, 3}, {1} }

Here, for elasticity values lower than −0.5, production and profit change their behavior. It is possible to delimit two distinct regions in the graphs:

- After −0.5, production and corresponding profits behave equally for both agents, and unitary prices rise with a low speed. It is a context of full supply.
- Below −0.5, agent {1} is not able to produce additional energy, and unitary prices rise at higher rates, as a consequence of the shortage in production ability. So, players experience rising profits until reaching the corresponding production limits.

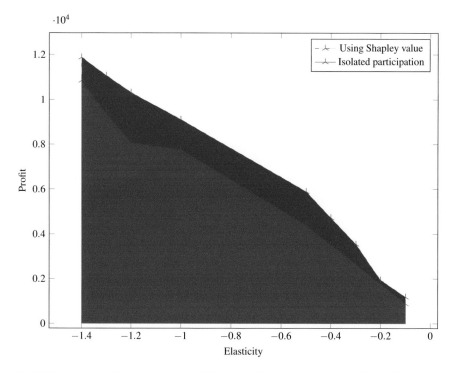

Fig. 7.17 Isolated profit and assigment with Shapley value when in grand coalition—Firm 1

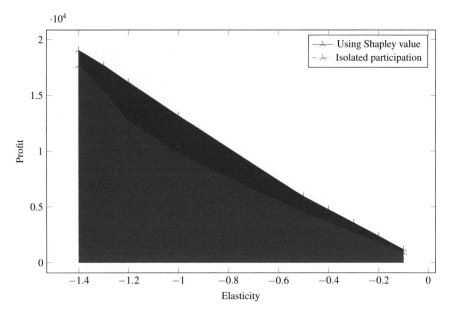

Fig. 7.18 Isolated profit and assigment with Shapley value when in grand coalition—Firm 2

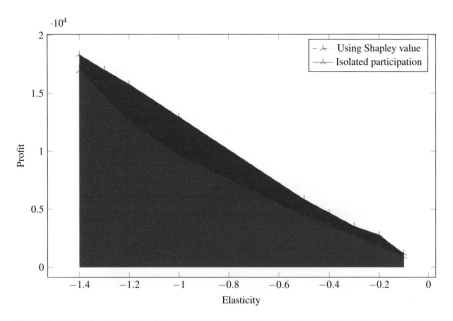

Fig. 7.19 Isolated profit and assigment with Shapley value when in grand coalition—Firm 3

7.3.5   Grand coalition {{1, 2, 3 }}

This distinguished configuration presents equilibrium solutions with (almost) linear behavior, what may be attributed to the lack of competition. This is so because there is only one apparent agent furnishing energy, although physically three firms sustain its operation, working cooperatively.

7.3.6   Interpretation of Obtained Results

Figures 7.17, 7.18 and 7.19 allow us to conclude about coalition formation convenience for each firm as a function of elasticity behavior. A conservative viewpoint is being adopted, considering that the comparison is made between the profits obtained in a configuration of isolated participation (a relatively unfavorable circumstance) and the Shapley values computed from the grand coalition, typically a convenient situation, observing the high values of unitary prices associated to the several elasticity values.

The graphs show that in $[-1.3, -0.3]$ the three firms have more profitable results when acting in a consortium (grand coalition). On the other hand, outside that interval there is a propensity to invert that trend, that is, for elasticty values inferior to -1.4 or greater than -0.1, the obtained results indicate the existence of a counterexample for the general belief that larger profits always correspond to the formation of coalitions.

7.4   Conclusion

The chapter presented an approach for investigation of the influence of price-demand elasticity in electricity markets regarding the profitability of coalition formation among firms. The method used in the solution is based on facing each auction instance as a GNEP, converting the original problem into a constrained global optimization problem, and discovering at least one global minimizer, each of which corresponds to a generalized Nash equilibrium—points that represent stable configurations at which agents cannot improve their positions in a unilateral way. Therefore, it becomes possible to evaluate how interesting is to join and cooperate in order to get bigger advantages than those possible when acting in an isolated manner. This may be done by varying the relationship between unitary price and total energy demand, and finding the solutions for the resulting GNEP's—each Nash equilibrium shows the profits gained by existing players, be them coalitions or real firms.

In order to find the profits corresponding to each firm, the concept of bilateral Shapley value is used, making it possible to estimate how profitable would it be to form coalitions under diverse market conditions. Although the proposed method is very general, it was exemplified here by means of a very meaningful problem, previously presented in [3]. The presented example made it possible to assess the importance of price-demand elasticity, demonstrating that its variation may give rise to different sets of coalitions among generating firms, and the resulting profits may be attributed by means of concepts such as the bilateral Shapley value.

Also, it was shown that it is possible, for a general purpose metaheuristic optimization method, to find solutions for GNEP's in a single optimization step. So, a major target is to demonstrate the efficacy of the Fuzzy ASA method with respect to its ability to find solutions for GNEP's without using multi-level optimization techniques. To get there, the optimization formulation of the problem of computing generalized Nash equilibria studied in the literature [12–14] was changed, meaning that the original task is transformed into a single constrained optimization problem by means of a penalty method approach. Hence, considering the wide applicability of Fuzzy ASA, the presented method may be more general than some of the already existing ones in many circumstances, observing that Fuzzy ASA is able to deal with nondifferentiable functions defined on nonconvex domains. Consequently, the main point here is to offer an effective tool to evaluate and synthesize economic strategies capable of coping with significant problems in the electric energy market.

References

1. Aguiar e Oliveira Jr, H.: Coalition formation feasibility and Nash-Cournot equilibrium problems in electricity markets: a fuzzy ASA approach. Appl. Soft Comput. **35**(12), 1 (2015)
2. Chalkiadakis, G., Elkind, E., Wooldridge, M.: Computational Aspects of Cooperative Game Theory. Morgan & Claypool, San Rafael (2012)
3. Contreras, J., Klusch, M., Krawczyk, J.B.: Numerical solutions to Nash-Cournot equilibria in coupled constraint electricity markets. IEEE Trans. Power Syst. **19**, 195–206 (2004)
4. Hobbs, B.F.: Linear complementarity models of Nash-Cournot competition in bilateral and POOLCO power markets. IEEE Trans. Power Syst. **16**, 194–202 (2001)
5. Krawczyk, J.B., Uryasev, S.: Relaxation algorithms to find Nash equilibria with economic applications. Environ. Model. Assess. **5**, 63–73 (2000)
6. Mastroeni, G.: Gap functions for equilibrium problems. J. Glob. Opt. **27**, 411–426 (2003)
7. Ongsakul, W., Vo, D.N.: Artificial Intelligence in Power System Optimization. CRC Press, Boca Raton (2013)
8. Peters, H.: Game Theory - A Multi-Leveled Approach. Springer, Berlin (2008)
9. Rosen, J.B.: Existence and uniqueness of equilibrium points for concave N-person games. Econometrica **33**, 520–534 (1965)
10. Shapley, L.S.: A value for n-person games. In: Tucker, A.W., Kuhn, H.W. (eds.) Contributions to the Theory of Games II, pp. 307–317. Princeton University Press, Princeton (1953)
11. Uryasev, S., Rubinstein, R.Y.: On relaxation algorithms in computation of noncooperative equilibria. IEEE Trans. Autom. Control **39**, 1263–1267 (1994)
12. von Heusinger, A., Kanzow, C.: Optimization reformulations of the generalized Nash equilibrium problem using Nikaido-Isoda type functions. Preprint 269. University of Würzburg, Würzburg, Germany, Institute of Mathematics (2006)
13. von Heusinger, A., Kanzow, C.: SC^1 optimization reformulations of the generalized Nash equilibrium problem. Opt. Methods Softw. **23**, 953–973 (2008)
14. von Heusinger, A., Kanzow, C., Fukushima, M.: Newton's method for computing a normalized equilibrium in the generalized Nash game through fixed point formulation, Preprint 286. University of Würzburg, Würzburg, Germany, Institute of Mathematics (2009)

Chapter 8
Epilogue

Abstract This short chapter aims to summarize the content and intentions of the book.

8.1 Final Considerations

When writing this book one major intention was to demonstrate once more the enormous importance of global optimization techniques and their applications to the realm of Applied Mathematics. This is so because they make it possible to attain practical results even in the absence of theoretical facts about a given problem, be it related to optimization-based engineering design or the search for Nash equilibria in economic systems.

Another very important goal of the book is to show that, by applying results of Topology, it is possible to extend the scope of current evolutionary techniques to manifolds, what is very positive, taking into account that besides improving the search process in constrained optimization problems, it is also possible to extend the scope of present metaheuristic methods with minor adaptations, if any. These extensions are compatible with the fact that all mathematical methods need formal proof to be validated, and this directive must be respected in order to not get trapped inside dangerous intellectual "black holes". So, considering that many well-known metaheuristic techniques already have firm theoretical foundations justifying their dynamics, the enlargement of their scope presented here offers a great opportunity for obtaining new results without significant additional effort, because the transition to manifolds is relatively simple. Also, the chapters devoted to the solution of generalized Nash equilibrium problems and their applications are a kind of invitation for finding new and useful applications of this powerful concept, mainly when using optimization based methods.

© Springer International Publishing Switzerland 2016
H. Aguiar e Oliveira Junior, *Evolutionary Global Optimization,*
Manifolds and Applications, Studies in Systems, Decision and Control 43,
DOI 10.1007/978-3-319-26467-7_8

Finally, permeating the whole book is the Fuzzy Adaptive Simulated Annealing paradigm, that exhibits a lot of potential to solve many different kinds of complex problems, when it is possible to build adequate objective functions that may completely portray all necessary conditions for each particular design task.

Index

© Springer International Publishing Switzerland 2016
H. Aguiar e Oliveira Junior, *Evolutionary Global Optimization,
Manifolds and Applications*, Studies in Systems, Decision and Control 43,
DOI 10.1007/978-3-319-26467-7

Printed in the United States
By Bookmasters